WHEN YOU OPEN A BOOK, YOU OPEN THE MIND OF THE AUTHOR.

The Smart Guy, the Intelligent Guy, and the Wise Monkey

Asterism principia

The eye that sees clearly in the midst of darkness.
The language of the anonymous cosmos.
Simplicity.

Teni A.

Paperback Edition ISBN
978-0-9916306-3-9

Printed in the United States of America.
8763-09 13 01 16 CM: 097-342; 181833
BC courtesy—TATI316
000726/A113-2022

Second Edition.
Published 2023

The ideas in this book are products of the author's thoughts, and any resemblance to actual occurrences, concepts, locations, or people, alive or deceased, is completely coincidental.
All Bible verses quoted in this publication are from the King James Version. The majority of the definitions in this book are the author's opinion and are indicated with "as I defined it." Some definitions used in this publication are quoted from the *Oxford Learner's Dictionary and Oxford Languages*.

Insanity is doing the same thing over and over again and expecting to achieve different results.
If you can't explain it simply, you don't understand it well enough. –Albert Einstein

Contents

To My Skeptical Mind

The distance between life and death is a question of time.

Time, time, and time.

The life of the unborn is the hope of the dead.

It's the duty of life to open the door. It's the duty of death to close it.

I don't hate or love either life or death. I respect them.

A moment of truth can swallow a thousand years of lies.

One ounce of courage is heavier than a thousand pounds of fear.

Night after night, I sit alone in the dark. I don't know if I should open or close my eyes. What difference will it make?

As little as the eye is, it can see the world but it can't see itself.

No melody is more enlightening to my ears than the national anthem of a sovereign nation, for they are composed in times when humans have chosen freedom or death.

Show me a person who is not afraid to die, and I'll show you a person who can accomplish the unthinkable.

I've been in the world for quite some time, and I'd be lying if I said I hadn't seen, learned, experienced, or discovered something.

I guess as a human full of imperfections, I'm supposed to be afraid to voice my opinion about the world I'll never fully understand, huh? I apologize if I'm not.

I can't conclusively say if I'm wrong or right. What I do know is, I'm neither.

Whether or not one or two people disagree with me doesn't matter. I just hope that after they read this book, I've been able to move them one step closer to a full understanding of the world we live in.

So, dear skeptic mind, trust me once and don't trust me again.

Introduction

I'm not the type of author who will bombard you with a bunch of logical yet hypothetical proofs of a deity. I say this with humility and a sense of humor.

In this book, I try to answer some of the toughest questions I've come across in physics and religious studies.

Using new theories derived from known observations, reasoning, critical thinking, and common analogies—the majority of which you will probably be familiar with—I present my answers in basic mathematics and science languages. Thus, all I ask you to do is *think* as I connect the dots. For every existing claim or theory I disagree with, I'll offer what I believe is a more reasonable alternative.

In order to answer some fundamental questions in physics, I approach the basis of quantum theory from a different but familiar perspective. Instead of taking you through abstract mathematics that won't explain the meaning of reality, I tackle the questions you often ask and in response are told to shut up and calculate. A response which I found somewhat funny and unintendedly challenging.

Not everyone will agree with my ideas, but whether you do or not, when you finish reading this book, it will be difficult for you *not* to think about what you have read.

In this book, you will come across statements that may not resemble what you think, know, or believe. Before you conclude that these statements are wrong, what's presented in this book are theories of my own. In addition, to make sure my opinion is as easy to understand as possible, I intentionally made sure this book wasn't too long or complex. Thus, it may not contain my entire theories and arguments.

What is this?

Everything in existence—including what you believe and don't believe—can be categorized in one of two ways.

(Category one)		(Category two)
Invisible	or	Visible
Cause		Effect
Energy		Matter
Creator		Creature
Designer		Design
Occupation		Occupant
Non-Created Thing		Created Thing

Invisible, as I define it, is something the eyes can't see without help from something greater.

Proof, as I define it, is evidence, either in the nature of its presentation or a link to the actual occurrence that cannot be subjected to any reasonable opposing argument.

Before any reasonable argument starts, the party bringing forward the argument must first present a clear definition of the thing in question. So, for the sake of clarification and simplicity, let us refer to every word in Category one as "God" or "characteristics of God," and every word in Category two as "everything." When I use the word "God," please refer to any of the characteristics in Category one to understand what I mean.

The difference between what we *know*, *think*, or *believe*, as these terms are used in this book, is that proofs support what we know, while beliefs and thoughts are often based on reasoning, not concrete evidence.

What is this?

There are three laws of existence that simplify the complex relationship between God and everything. Right now, you don't have to agree with them, and whatever questions you have will be answered by the end of this book.

The proper definition of *creature* and *creator* is presented in chapter six. However, "God" is the Creator and beginning referred to in the first and second laws of existence. This beginning is the same as "everything," yet may differ in magnitude and appearance.

First Law of Existence

For every creature, there's a beginning, and if every beginning must have a beginning, there would be no beginning. Therefore, for the first creature to exist, there must have been a beginning that had no beginning and wasn't created.

Second Law of Existence

If a creature or design had no creator or designer, it would not exist; therefore, it would not be created or designed.

The third law of existence is discussed in the last chapter of this book.

Chapter 1

Space and Empty Space

As a child in a religious home, I went to church more than the pope. Attending weekly church services wasn't what bored me as a youngster. What baffled me were the teachings in the Scriptures. How did Jesus walk on water? How did a serpent talk to Eve in the garden? How do our offerings get to heaven? And where are heaven and hell located? It didn't matter who I turned to for answers; everyone told me the same thing: "You will understand when you grow up."

But as I got older, these questions only became tougher. The first chapter of the Bible says, "In the beginning God created the heaven and the earth" (Genesis 1:1). It is also written, "Thus saith the Lord, The heaven is my throne, and the earth is my footstool: where is the house that ye build unto me? And where is the place of my rest?" (Isaiah 66:1).

I often wondered if God created heaven, what space did he occupy before heaven was created? If you think about this question with an honest and open mind, you will come to one rational conclusion: either an occupation existed before God, or God himself *is* an occupation. Ask yourself, could the existence of aquatic animals precede the existence of water? This is where my journey into critical thinking began.

Does space have a beginning?

An honest attempt to answer this question would lead to a deciding question: what is the fabric of space made of? In order to provide a rational answer to this question, one needs to first agree or disagree with the logic in the theory that describes how the universe was created. This initial argument would resemble scrutiny, but it's not meant to refute the big bang theory . . . yet. Rather, it is intended to create reasonable doubt about the origin of space.

It's important to first explain the difference between *space* and *empty space*. Space might not be empty, yet the property that fills the area can allow the occupation of another form of property. For instance, the Atlantic Ocean and the interior of an inflated balloon could be referred to as space. Empty space is an area containing absolutely nothing; no form of matter or energy exists there.

But is there anywhere in the universe that is empty? Is 99.9 percent of the universe, including areas within the atom, empty space? Well, that's what most people thought until the study of quantum physics revealed that virtual particles could emerge from a place that was once considered a pure vacuum.

The big bang theory

In 1931, a Belgium physicist named Georges Lemaître (1894–1966) published a hypothesis where he claimed the universe (space and time) came into existence from the explosion of a single primeval atom. Fred Hoyle, a physicist who opposed the idea at the time, jokingly called it a "big bang" during a radio broadcast. Since then, this hypothesis formulated by Lemaître that tries to explain the origin of the universe has been named the "big bang theory." Or as many physicists refers to it, "a tiny hot ball of matter that began the universe."

According to physicist Stephen Hawking (1942–2018) in his book *A Brief History of Time*, the early universe was a single point of infinitely small, massively dense space-time. Roughly around 13.8 billion years ago, for some unknown reason, this tiny atom began to expand, giving rise to the particles that form all the matter we see today. According to the big bang theory, time and space, or empty space, did not exist before the big bang.

Let me state a familiar observation. As far as I could demonstrate, and based on the definition of matter, "an occupant whose existence is dependent on its occupation cannot exist before its occupation." That being said, all things in existence can be classified as either an occupation or an occupant. A primeval atom, regardless of its size and description, is classified as a "thing." Hence, the big bang couldn't have created "everything." If you don't agree with that, at least you can agree that the big bang didn't create itself.

Another claim that often raises questions is the notion that the big bang didn't expand inside anything; instead, this tiny, hot ball of matter expanded and stretched space. It seems like almost anyone who dares to question the origin of the big bang today is often labeled as not understanding the theory.

If a teacher claims a student doesn't understand the origin of the big bang but the teacher cannot explain it, how do they know the student is wrong?

If someone who opposes the origin of the universe is labeled as not understanding the big bang theory, yet there's no publication where the theory is well defined with specifics by the founder, I have no choice but to believe that the big bang theory was created to intentionally mislead.

What well-respected physicists think and publish about an idea is not the same thing as an indisputable definition of the idea by the author. As far as I understand the big bang concept, around 13.8 billion years ago, a *primeval atom* disintegrated in an explosion, and space expanded outward at nearly the speed of light. Before this event occurred, space and time did not exist.

Any argument contradicting this popular claim would make the big bang theory similar to religious writings. Without wishing to offend anyone, we know the writings in religious books are not properly explained. This is why pastors can take a verse from the Bible and interpret it in different ways. But have you ever heard any biologist teach the theory of evolution contrary to what Charles Darwin published during his lifetime? I haven't. What specifically does the big bang theory state? And most importantly, according to whom?

Georges Lemaître referred to the beginning of the universe as the "explosion of a single primeval atom." Many physicists today would claim it was not an explosion, but an expansion of space. Stephen Hawking argued that the universe began as an infinitely dense point called a *singularity*. Whenever there's an argument over what a word means, we refer to the dictionary for a complete definition and origin. We do the same with constitutions. It would have been better to have an in-depth definition formulated by the first proposer (Georges Lemaître) that we could use to settle disputes over what the big bang theory really states. For example, is the tiny, hot ball of matter an atom with one or more protons? Is the singularity the size of a quark or infinitely smaller than an electron?

Regardless of size, I don't have to explain how it is logically impossible for the primeval atom to create itself. On the other hand, most people believe that it created empty space. How is this possible? Instincts are not always right, yet many people would agree that the first step toward solving a problem is to follow your instincts.

Instincts, as I define them, are unproven conclusions made by the mind or body of a life-form that allows it to respond directly to its environment. Briefly, let's visualize how the big bang theory explained the origin of the universe, first with our instincts, followed by logic, and then questions.

Visual explanation of the big bang

In case you are not familiar with the big bang theory, here is how it starts: in the beginning, there was nothing, and all of a sudden, an infinitely tiny ball of matter began to expand, stretching out space and . . . stop right there. With all due respect, how did matter and space expand outward if there was no outside for them to expand into?

Empty space can *only* be created by removing something from an occupied area. Just as darkness represents the absence of visible light, empty space represents the absence of something. Any claim or theory that says otherwise cannot withstand a testable and demonstrable protocol. Since the big bang is not an irrefutable occurrence, the theory cannot be presented as evidence.

An *irrefutable occurrence* is something whose state of existence cannot be disputed even though the evidence for or against it may be right or wrong. For instance, we could argue against the theory of an invisible hand moving planets around the sun, but we can't dispute orbits. A defense attorney can argue that his client didn't kill Paul, but he can't dispute that Paul's dead body is at the morgue. In this example, orbits and the dead body are irrefutable occurrences.

It is evident that Paul is dead from the gunshot wound to his head; that's why we are looking for evidence to identify the killer.

Initially, before humankind started practicing science, no one thought the world had a beginning, since there was no irrefutable occurrence to suggest it. But based on religious teachings, which say that God created heaven and the earth, we *believe* the world, which is later referred to as the *universe,* had a beginning. That's why scientists instinctively started looking for evidence to support a created universe. And in fact, we found one—and its origin is as difficult to explain as the origin of the biblical God himself. In fact, in the times before Christ, the religious belief was so strong that people believed the stars in the night sky were the only things in heaven and the earth was the center of our universe.

Questions about the big bang

Since the majority of evidence presented for the big bang theory is based on reasoning, I implore you to consider these questions:

- According to current cosmology, space is expanding. But assume for a moment that space stops expanding. Would a particle of matter that began accelerating at the moment of the big bang run into a space wall? If not, why does space have to expand if the motion of matter through space cannot be stopped by the intangible fabric of space?

- Before the big bang, how would you describe the absence of empty space around the primeval atom? In other words, would you say the primeval atom was a point where motion was impossible due to the absence of empty space?
- If space or empty space didn't exist before the big bang, how do we reasonably describe where the expansion happened?
- Would you define the hypothetical primeval atom that began the universe as the *only* matter that had mass yet didn't occupy space?
- If you believe space started from an infinitesimally small point, why couldn't space have started as a large place?
- What does the fabric of space consist of?
- If, instead of becoming infinitely big as we know it today, space had stopped expanding when it had a diameter the size of earth, how would you describe the edge of space?

If the first law of existence I proposed earlier doesn't sit well with you, here are a few more questions to ponder:

1. Where did the big bang come from?

2. What created the big bang?
3. And if the big bang marked the beginning of our universe, how did it begin?

I've learned from personal experience that it is not cool to ask any of these questions. But there's nothing wrong with challenging a Christian to explain what or who created the biblical God.

To be fair to mankind, I think we need to ask anyone who believes the universe came into existence through a cause to present a reasonable explanation of how it came into existence in the first place.

Insights from physics and reasoning

Without irrefutable evidence of the big bang's origin, it seems to me that the reason for describing the ball of matter that predated it as infinitesimally small is to avoid explaining how the big bang came into existence. If all the space between the subatomic particles that make up the sun's atoms were removed, the sun would shrink to the size of Jupiter or smaller.

On the other hand, based on the size of the smallest star we know, I doubt if *all* the matter in the entire universe today could shrink to the size of an atom or proton, as hypothesized by the big bang theory.

When discussing the origin of the big bang, we spend most of the time emphasizing the intricacy of the singularity or primeval atom and how infinitely tiny and dense it was before the expansion or explosion took place (as if any of us were there when it happened).

According to our current understanding of physics, the smallest things in the universe are fermions and bosons, but no one bothers to ask if the primeval atom should be described as a fermion, boson, hadron, or just atom as first described by Georges Lemaître. (This question is based on what we *know*, not what we *think* or *believe*.)

Irrespective of how infinitesimally small we try to make the atom before the explosion, it can't be described as a *fermion* (an indivisible particle). I mean, this is just a no-brainer for a physicist. Fermions don't explode, nor can they be divided further. That's why we call them elementary particles.

In this sense, the big bang has to be some sort of hadron or atom. Look at it this way: Do you believe the tiny stuff that once contained *all* the matter in the universe was massless and could not be divided further? If not, then let's refer back to the definition of matter. Matter is *anything* with mass and *occupying space*. Hence, isn't it reasonable to ask what space the primeval atom *occupied* before expanding or exploding?

I'm aware of the hypothetical claim that the big bang happened everywhere at the same time. The ultimate question in this argument is what came before? *Before* the big bang happened, how do we describe the *total absence* of empty space?

If your answer is that nothing existed before the big bang, it's important to understand that you can't reasonably demonstrate empty space without invoking or removing some sort of obstructing property. Therefore, in theory, empty space is actually another way of defining or describing absolutely nothing in an area.

Trying to argue against this reasoning would be like saying, "Let's disregard everything we know to be true in favor of the big bang theory." And that's exactly the unintentional attitude we are living through right now.

Think about it. In the beginning, if there was absolutely nothing around the tiny, hot ball of matter, the absence of something should make you wonder why the big bang needed to create empty space. You create empty space by moving *something* out of the way, right? If there was *nothing* to move out of the way, it means empty space already existed.

The thought that empty space was created arises from an early religious belief that God created *everything* in existence; therefore, Georges Lemaître must have felt it was necessary for it to include the creation of space.

Try this thought scenario to understand how the creation of empty space is illogical. The entire universe means energy, matter, and space, correct? In your imagination, remove empty space from the universe. Do you notice how unreasonable it would sound to say that without space, matter or energy would exist alone? I'm sure your brain would ask, "Alone where?" Therefore, when someone mentions the big bang "expansion" or "explosion," isn't it reasonable to ask them *where* it happened? I believe it does. This is why the definition of matter is not complete until a body with mass occupies space.

Cosmic microwave background

One of the discoveries of the twentieth century was the existence of cosmic microwave background (CMB). It was discovered by two physicists, Arno Penzias and Robert Wilson, in 1963.

According to the big bang theory, CMB is electromagnetic radiation left over by the big bang explosion or expansion. As this finding is credible to me, it leaves behind a question. Anyone who understands how heat leaves behind this kind of radiation, such as after you warm your food in a microwave, would agree that cosmic microwave background only shows a telltale sign of a once hot or highly radiated cosmic atmosphere and is not necessarily related to a created universe or some kind of explosion.

If the big bang were a defendant on trial, the knee-jerk conclusion that CMB resulted from it would never hold up in a court of law. Anyone willing to reason with different possibilities knows that it wasn't as if we found DNA or fingerprints in CMB linking it directly to a specific cause.

There's no direct proof that linked the CMB to an explosion or expansion. If Lemaître believes the primeval atom started with a high temperature, another theory of a non-created universe could argue that the CMB is an indelible property of space-time that formed over a long period of quantum fluctuations, and not necessarily one that was left behind after an explosion. For instance, because I made a prediction that deadly radiation will be found in a city if truly a nuclear bomb was detonated there in the past, doesn't necessarily mean if deadly radiation is found in the city, it must have resulted from the detonation of a nuclear bomb. The deadly radiation found in the city could have also resulted from a nuclear plant meltdown. As a result, because the big bang is not an irrefutable occurrence, the notion that the CMB resulted from it is a reasonable fallacy.

Cosmic background radiation can be compared to a white van that had been used during a home invasion. If two tourists walking in a park stumble upon an abandoned white van and nobody in town knows how the van got there, they immediately jump to the conclusion that this was the same van used during the home invasion. Even though it's a white van, their assumption wouldn't necessarily be right.

Expanding universe

If I claim the universe was created 13.8 billion years ago from the violent explosion of an extremely hot and compressed ball of matter the size of a baseball that stretched out space and time, do you know I can also back up my claim with all the evidence currently being used to back up Lemaitre's "primeval atom" or Hawking's "singularity"? And any opposing argument you raise against my claim can also be used against the big bang theory. Keep in mind that the existence of the hypothetical "primeval atom," "singularity," and my "baseball-sized" big bang are not *irrefutable occurrences*.

If the expanding universe were put on trial, the last argument that should be brought up to defend the creation of empty space is the expansion of space.

Although the fabric of space has not yet been directly observed, it is not the same as the texture of matter or energy. But as long as we can only demonstrate stretching or expansion with a form of matter or energy, the principles that apply when matter is expanding must also apply to the expansion of space.

In order to visualize this concept, think about the expanding universe as a balloon. When inflating a balloon, it is obvious that an expansion can only occur with things that have an edge. When expansion occurs between two bodies lying on the same surface, either one body needs to be stationary while the other body moves away, or the bodies must be moving in opposite directions.

In the case of expansion that starts at a single point and spreads outward, it *must* be uniform in all directions.

One assertion that arose from the big bang theory is that expanding space is the reason for receding galaxies. Let me explain one of the contradictions in this claim with this simple demonstration. (This initial demonstration is performed in one-dimensional space.) Assume that the two Gs in parentheses are galaxies: (G1 and G2). Now imagine that the space between the two Gs is expanding.

The first way expansion can occur is if G1 is moving backward and G2 is moving forward (expansion method 1). The second way expansion can occur is if G1 is stationary while G2 is moving forward, and vice versa (expansion method 2).

Now apply either of the two expansion methods to concurrently expand the spaces between the Gs in these parentheses: (G1, G2, G3, G4, G5, G6, and G7). You will notice that unless you start the expansion by simultaneously pulling both parentheses in opposite directions, there will be an unreasonable nonuniform expansion of space between the galaxies.

It's easy to demonstrate the expanding universe with something that has an edge, like a balloon, cookie dough, or a rubber band, but when you question if the universe has an edge, the logic starts to break down.

If someone is trying to demonstrate the expansion of space with a piece of elastic, ask this question: Would the red and blue shifting of light waves be impossible if space is not expanding? I doubt it, but if the answer is yes, follow up with this question: Why don't we notice the attraction between every point of mass in the universe, as Newton's law of universal gravity states?

One of the most common answers to explain this is that gravity is *too weak*. I'm unbiased when I use this reply. Many of you have probably heard this statement numerous times before.

To demonstrate the expansion of space, let's continue with the balloon example. But instead of the dots commonly drawn to represent galaxies, draw a light wave and curved space (both with a *U* shape) on the balloon and inflate it. As it expands, the *U* shape gets bigger too. So why hasn't the expansion of space over a long period of time affected its curvature as it has the light waves? The common answer is "gravitational force." You mean the gravitational force that was deemed *too weak*?

Moreover, are orbits held together by a force of attraction *or* curvature in space? A reasonable teacher should be honest with you at this point.

In a world where life and liberty sometimes depend on scientific evidence and testimony, we can't afford to treat a scientific concept like dice. For instance, when describing a black hole, we use Newtonian gravity to explain why a star shrinks or collapses under its own gravitational pull. But after the star explodes, it leaves behind a core, and we use Einstein's gravity to describe a black hole as the warping of space so strong that it becomes an infinitely curved hole. From a realist's point of view, we are switching the concept of gravity back and forth depending on the circumstances we need it to fit.

I don't expect science to be perfect in all it does, but when our theories pretend to know the problem yet cannot solve it, it makes it difficult for the mind trying to solve the problem to know exactly where the problem lies.

I hope it hasn't become normal for humans to retain old ideas even if they have been proven wrong, such as the earth is flat or was created in six days, the latter of which many people today still believe is true.

Gravity

It's confusing how a majority of astronomers agree that gravity is a curvature of space, but when it comes to the expanding universe, they are quick to say that gravity is a force of attraction that is slowing down the expansion. We can't claim that gravity is too weak except when it comes to slowing down the universe's expansion. Or perhaps I missed the grand unification of both Newton's and Einstein's gravity.

I'll talk more about gravity in chapter three. However, based on the current understanding of physics, gravity is a weak force. Therefore, the notion that gravity is slowing down the expansion of the entire universe is unreasonable. That's like saying the movement of a goldfish can slow down a raging storm in the Pacific Ocean.

Am I claiming that space cannot undergo some sort of change? No. Am I claiming that the motion of galaxies is not uniform in all directions and therefore couldn't have started from a single point? Yes.

Initial conclusion

The age-old proposition that expansion means space and matter must have originated from an infinitely dense point called the big bang needs scrutiny. The first question that needs to be answered is whether in the beginning, matter was packed into an infinitely dense space or if space was inside an infinitely dense matter.

Let's assume both are woven together like an infinitely dense fabric in the middle of nowhere. How long are we going to continue creating hypotheses to defend logic that will never make sense to most people?

The part of the doubt I intend to establish in this chapter is that any theory trying to explain the beginning of an occupant but doesn't begin with an occupation will raise the following question: Where did it happen? Only by inserting an occupation can one reasonably answer it. And because the fabric of space is intangible in nature, it is unreasonable to human understanding that a tangible occupant created an intangible occupation.

There's an old saying that what doesn't have an end can't have a beginning, and vice versa. As far as we know, the part of the universe we refer to as space doesn't have an end. Consequently, space couldn't have had a beginning.

Chapter 2

The Language of the Universe: Mathematics and Science

We must admit that most of our earlier ancestors had little to no understanding of the world. Thus, we can't treat everything they believed as fact. Before the era of science and technology, what used to be known as ghosts are now being studied with instruments used to detect radiation. Faith was what we used when we had no telescope, but now that we have one—and not just any telescope, but the Hubble and Webb Space Telescope—we know the earth is not the center of the universe. However, as there are still numerous unsolved mysteries in the cosmos, telescopes will not get rid of faith.

The existence of God is, in part, lost inside religion. Today, when people are looking for God instead of studying nature and dissecting the visible things around them, they engage in religious studies that contain little to no factual basis. As I've come to realize, the greatest human mistake is drawing a line between God and science.

Due to the lack of clear definitions in religious books, early researchers started to refer to the study of God as science, and as a result of a hypothetical, faith-based notion that describes God as a mysterious being, religionists have unintentionally created another imaginary God to oppose science. It turns out that science isn't the epitome of evil established by humans to ridicule God; instead, science is discovering evidence of God that religion cannot prove.

I could relate this to my personal experience. As a juvenile, I knew my parents had sent me to school to learn, yet I didn't understand why I had to study English, biology, chemistry, physics, mathematics, art, and those other annoying subjects you hate when you are young. But as I got older and started to make friends, I began to meet people of different ethnicities who spoke different languages, most of which I did not understand.

One evening in 2007, I arrived in Paris for the first time, and I couldn't understand a word anyone at the airport was saying. I remember thinking how

I wished I could understand every language in the world. As a curious realist, it didn't take long for me to reason with this thought. What if education is a language I needed to learn to understand the mysterious world in which I live? Perhaps knowledge is hidden in a book and only those who read will find it.

My attitude toward education started to change when I began to look at physics, chemistry, biology, mathematics, economics, and geography as languages I needed to learn if I wanted to understand the world.

Mathematics is a language I'm still struggling to grasp; nonetheless, I somehow found a way to fall in love with it. Although most scientific findings may not match what is recorded in religious books, the flimsy gap between faith and science would finally merge if religions could only see math and science as "languages" that humans have created to understand God.

Words often make matters worse in both religion and science. How did the biblical God create everything visible and invisible if he is also part of the invisible? I can't wrap my mind around the concept of a God who created himself. It makes no sense to me, and I believe it doesn't make sense to you either. It only makes sense if we say God created all created things.

Mathematics is intelligence

How did God create the world? Was intelligence involved in the creation of matter? What does it mean to have intelligence?

Intelligence is defined as the ability to acquire and apply knowledge and skills. It may come as a surprise when I say mathematics is the ability of an intelligent being to learn, think, and acquire knowledge and skills. If you are not surprised by this, it means you understand how "intelligence" refers to the ability to apply some sort of calculus to solve a problem.

An easy way to better understand this is to remove the ability to calculate from an intelligent being. A doctor knows a tree is better than a brain-dead human.

Let me clarify this. Your brain doesn't have a speedometer or odometer. Even so, before you cross the road, you can tell instinctively if the vehicle coming toward you is moving quickly or slowly, and how far away it is from your position. You can calculate distance, speed, and time without the help of a device because you have intelligence.

The same is true for a car or airplane on autopilot.

Mathematics lies at the root of all software. To paraphrase, anything that can apply mathematics to solve a problem has demonstrated intelligence.

Mathematics is the science of numbers and shapes. Numbers, shapes, and signs are the basic languages of mathematics. This is why the first thing you learn in mathematics when you are a child is how to count. The second thing you learn is how to add, subtract, divide, and multiply numbers. Furthermore, arithmetic operations help one understand the outcome when numbers are manipulated. Out of all the arithmetic operations in mathematics, only addition and multiplication can increase a number, and I'm sure you know that multiplication is greater than addition.

Humans use two methods to understand the universe: mathematics and science. Thus, this book will only be reasonable and unbiased if my explanations are spoken in one of these general languages.

If I think about creation in the language of mathematics, the question will become: What is the mathematical logic behind the creation of matter? Mathematically speaking, creation is more like adding or multiplying one or two numbers together to form a higher number, or adding one brick on top of another to build a house. The physical representations of abstract numbers during creation could be bricks, rocks, pipes, elements, molecules, atoms, or subatomic particles.

As seen in a multiplication table, to increase a numeric value, it is allowed to multiply a number by itself. Moreover, a number in the universe includes that number and its opposite (i.e., the number one in the universe means *one* and *anti-one*). This is similar to saying that where there's light, there is also darkness.

Since subtracting or dividing a number by itself won't increase the number, we can conclude that God applied either addition or multiplication to the building blocks of matter during creation. Whether the creation of matter took five days or five million decades isn't important here. What we are interested in is the mathematical explanation of the method applied during the creation of matter.

Can we mathematically express how something visible can come out of what appears to be invisible? This is also the question of how something came out of what appears to be nothing. By studying how addition and multiplication increase numeric values, I believe we can answer this question.

The force of nature in mathematics

Force is *any* influence that can cause a body to undergo change. Change or motion creates time. As a result, time can only be set in motion by a force. Therefore, whatever faith or method you have accepted to explain the beginning of time, it is without argument that force is necessary to set time in motion.

To put it simply, no matter what perspective of morality you belong to, your first cause *must* be described as a force.

If all there was in the beginning was a fundamental force that wasn't created (God, spiritual forces, the big bang, Allah, Buddha, whatever you prefer to call it), the goal here is to apply the addition and multiplication operation to numbers until we find one that has the power to bring something new into existence. Then we can assign that number an arithmetic operation to represent a beginning that has no beginning and cannot be created, as seen in the first law of existence.

Because most people believe the visible things we see around us today came out of nothing, I will begin this demonstration with zero.

Zero plus zero equals zero, and zero multiplied by zero equals zero. The numerical values we begin with equal our results. This means we didn't get something new out of multiplication.

One plus one equals two, and one multiplied by one equals one. Notice that even though multiplication is greater than addition, when applied to one, multiplication didn't produce something greater than addition. The numerical values we began with are equal to or less than our results. This means something new didn't come out of multiplication.

Two plus two equals four, and two multiplied by two equals four. Again, notice that even though multiplication is greater than addition, when applied to two, both addition and multiplication produce an equal result. Like the previous number, the numerical values we started with equal our results. This means something new didn't come out of multiplication.

Three plus three equals six, and three multiplied by three equals nine. Here you can see that something new has been created when the multiplication operation is applied to the number three, as we end up with more than we started with.

With simple mathematics, which I call the "multiplicative power," we can learn this: multiplication has the power to bring something greater into existence, but the power of multiplication to produce something greater than addition cannot be revealed until you get to three.

Let me replace abstract numbers with a physical quantity to explain how the act of multiplication is different from addition. If there are two boxes, each containing three identical objects, and you are asked to add or multiply the objects in the boxes, you can't physically produce the results of multiplication in the same way you can the results of addition.

Here's what I mean. Let's say I have two buckets, each containing three oranges. I place the buckets in front of you and tell you to prove that three oranges plus three oranges equals six.

You grab three oranges from one bucket with your right hand and the other three with your left hand and count them all together. This is what you say: "One, two, three, four, five, six—three plus three is equal to six." Then you put the oranges back into their respective buckets.

Now I ask you to prove that three oranges multiplied by three oranges equals nine. You can't wait to get rid of my annoying presence, so without thinking, you quickly grab the oranges like you did before—three in your right palm and the other three in your left. As you are about to start counting, you suddenly realize it will be impossible to produce nine oranges if you only have six. Yet you can't disprove the fact that three oranges multiplied by three oranges equals nine oranges.

A curious mathematician reading this could argue that three of the oranges will need to be counted twice to get to nine, but in theory, this means that unless the result is available, multiplication is not possible. This would be true if the physical result of multiplication could not be proven outside of mathematics. This is one of the ideas of this book.

The mathematical demonstration with the oranges explains why multiplication is necessary to bring something new into existence. Here is how three, or any number greater than three, becomes the multiplicative power: starting from zero, if you add and multiply a number by itself, you can't prove that multiplication is greater than addition until you get to three.

Based on this fact, mathematically speaking, I can assert that the characteristics of the fundamental force necessary to bring something new into existence out of what appears to be nothing, *must* be related to the numeric value "three" or a number greater than three.

The significance of the multiplicative power, not only as a characteristic found in nature but the role it plays in creation, can only be understood properly with the language of science. However, before we go into science, let's look at the application of the multiplicative power in religious beliefs.

Chapter 3

Multiplicative Power in Religion

I'll be asking you a question repeatedly in this chapter because I want you to *think*.

For those who may be wondering if I can back up my previous claim with religious studies, let's look at religion momentarily before I round things up with science. I'll be drawing my first references from the Bible, since it is the holy book of the largest religion today.

The first time God appeared in the form of a human was in Genesis 18. "And he lift up his eyes and looked, and, lo, three men stood by him: and when he saw them, he ran to meet them from the tent door, and bowed himself toward the ground" (Genesis 18:2). Out of all the numbers in mathematics, why did God come in three persons?

As a matter of fact, numerous times in the Scriptures, God is described as a Trinity. Out of all the numbers in mathematics, why is God described as three entities known as God the Father, God the Son, and God the Holy Spirit?

Most Christians believe that the second time God came in the form of a human is in the New Testament. I can't prove or disprove that Jesus Christ of Nazareth actually lived or is just a fictional character. However, something about his life and behavior is consistent with multiplicative power.

Christians believe Jesus Christ of Nazareth is Lord. During the life of Jesus, seven noteworthy things happened to him. First was his birth. There were many wise men in Bethlehem when Jesus was born, but only three brought a gift to Jesus. Out of all the numbers in mathematics, why three?

The second was the transfiguration. Jesus had twelve disciples during his life, but when he went up to the mountain for his transfiguration, he only took three disciples with him. Out of all the numbers in mathematics, why three?

The third was the temptation. If you only know one thing about Satan, you know the devil is very persistent. Yet Satan only tempted Jesus three times. Out of all the numbers in mathematics, why three?

The fourth was the Last Supper. Matthew 26:33–34 reads, "Peter answered and said unto him, Though all men shall be offended because of thee, yet will I never be offended. Jesus said unto him, Verily I say unto thee, That this night, before the cock crow, thou shalt deny me thrice." Out of all the numbers in mathematics, why three?

The fifth was the crucifixion. Golgotha, where Jesus was crucified, had no law restricting how many people could be crucified in a day, but when Jesus was crucified, there were only three crosses. Out of all the numbers in mathematics, why three?

The sixth was the resurrection. Jesus resurrected on the third day. If Jesus wanted to make sure no one had any doubts that he was the Messiah, he could have waited at least a month before he resurrected. Out of all the numbers in mathematics, why three?

Soon after his resurrection, Jesus had another feast with his disciples where he asked Peter *three times* if he loved him. Out of all the numbers in mathematics, why three?

The seventh was the ministry. A child is considered an adult upon reaching the age of eighteen. Even so, Jesus waited until he was thirty (three decades) before he began his full ministry. Out of all the numbers in mathematics, why three?

Jesus's ministry only lasted three years. He died at the age of thirty-three. Out of all the numbers in mathematics, why three?

In the Scriptures, there are thousands, if not millions, of angels in heaven. But in Revelation 14, three angels come out of heaven twice to announce the return of Jesus. Out of all the numbers in mathematics, why three?

I don't think the decision to choose or use a specific number is just a mere coincidence to someone as intelligent as Jesus, the "all-knowing Son of God." I believe Jesus knew *exactly* what he meant in John 2:19, where it is written, "Jesus answered and said unto them, Destroy this temple, and in three days I will raise it up." Why three, and not one or two days?

This three character and behavior isn't limited to Christianity alone. If you are a Muslim, you must have learned from the Quran that Allah hears and sees everything. But have you noticed that the prophet Muhammad (Peace be upon him) recited his du'a three times and he advised Muslims to say it at least three times when requesting something from Allah? Why does the number of repeating du'a start at three before Allah would acknowledge the prayer?

If you are a Hindu, I believe you've heard about the three main Hindu gods: Brahma the Creator; Vishnu the Preserver; and Shiva the Destroyer.

In Buddhism, Buddha is represented in three bodies called "the three bodies of the Buddha." Buddha could have existed in one or two forms and his followers would still worship him. But the entity chose to exist in three bodies.

Why do the main gods in all major religions come in three entities? The answer is obvious if you translate it into mathematics: the multiplicative power.

Ancient Greeks represented the Trinity (multiplicative power) with a triangle and an eye, meant to represent divine providence. The great Egyptians understood this in a not so mathematical sense—so much so that they built a triangular pyramid to honor the spirits of their dead pharaohs.

Even today, nearly all the great nations of the world use the triangle sign as a symbol of power. In fact, every government in the world is founded upon multiplicative power.

In high school, you learned that the US government is divided into three branches: legislative, executive, and judiciary.

Have you ever wondered why witches seem to come in threes? Well, look no further. According to King Solomon in Ecclesiastes 4:12, "And if one prevail against him, two shall withstand him; and a threefold cord is not quickly broken."

King Solomon is considered by many as the wisest man to ever live. The majority of his writings are used and studied today by many who practice magic and various non-secular beliefs. However, as rich as religious books are in wisdom, they cannot be used to explain reality in a practical or demonstrable sense. Thus, even though I had explained the multiplicative power in terms of mathematics, I have to use the language of science to explain the application of the multiplicative power in every aspect of reality.

Chapter 4

Multiplicative Power in the Universe

Our understanding of the world is based on what we perceive with our senses. This is why most people won't believe in something unless it is proven or perhaps explained in a reasonable language. I must confess, I'm one of those people. That's not to say I don't believe in spirituality. I'm just skeptical about the idea of an omnipresent God whose existence cannot be proven. My skepticism sometimes got me looking for clues in the Scriptures. Four passages drew my attention to science:

I. Can any hide himself in secret places that I shall not see him? Saith the LORD. Do not I fill heaven and earth? Saith the LORD. (Jeremiah 23:24)

II. Through faith we understand that the worlds were framed by the word of God, so that things which are seen were not made of things which do appear. (Hebrews 11:3)

III. For in him we live, and move, and have our being; as certain also of your own poets have said, For we are also his offspring. (Acts 17:28)

IV. For the invisible things of him from the creation of the world are clearly seen, being understood by the things that are made, even his eternal power and Godhead; so that they are without excuse. (Romans 1:20)

I find it hard to believe that these four passages are speaking in parables, especially when they are obviously talking about something verifiable. Where the Scriptures talk about blessings, lands, and tithes, humans don't refer to those as parables. But when it comes to the existence of God or something we can't explain, we are quick to introduce the use of faith.

Before the use of the telescope, faith, just like abstract mathematics, is what our early ancestors used to make sense of reality. The difference between them is that mathematics can be demonstrated in the manipulations of writing and drawing, while faith can only be demonstrated in thoughts and words. However, if thoughts and words are convincing enough, it could cause one to begin to believe it is possible in reality.

Likewise, if the manipulations of abstract writings, such as numbers and shapes, are convincing enough, it could cause one to begin to believe it is possible in reality.

Instead of using faith, I believe if I apply mathematics and science to the four Scriptures above, it is possible to discover the true nature of God and everything. Romans 1:20, in particular, caused me to conclude that if God wasn't wearing a glove during the creation of the world, then a clue of his intelligent nature must be present in all the foundations of existence. To put this another way, just as a child possesses the DNA of their parents, God's intelligent nature must be present in all created and non-created things.

The universe

The universe, as far as we know it, is made up of three things: energy, matter, and space. The motion of matter in relation to space is known as *time.*

Time is thought to be one of the greatest mysteries of the universe. It has often been a topic of controversy among scientists, especially when dealing with Einstein's general theory of relativity. Despite the controversies surrounding the nature of time, one thing is agreed upon by the majority: time is divided into three parts—past, present, and future.

Another aspect that continues to baffle us is the nature of space. Despite all the arguments surrounding the shape of space, it is universally understood that it has three dimensions: height, length, and width.

The third greatest mystery in the early days of science was matter, which is anything that has mass and occupies space.

In the early nineteenth century, long before the idea of the big bang was introduced, legend has it that some curious researchers took a stone and began to break it into smaller parts. They broke the smallest part down to a point where they could not further divide it. This smallest unit of matter was dubbed "the atom" by a scientist named John Dalton (1766–1844). At the time, it was agreed that the atom was the building block of all visible matter.

The celebration only lasted until the late nineteenth century, when a scientist named Joseph John Thompson (1856–1940) discovered the electron.

Later discovery revealed there are three kinds of these particles—electrons, muons, and tauons.

In the early twentieth century, two more scientists emerged with the discovery of the proton and neutron. It turns out that the atom, which everybody thought was indivisible, was made up of three things: protons, neutrons, and electrons.

Like previous discoveries, the excitement didn't last long. As research continued, things only got more interesting. In 1964, two physicists, Murray Gell-Mann and George Zweig, who were working on the theory of strong interaction symmetry, proposed the existence of subatomic particles known as *quarks*.

Four years later, scientists working on inelastic scattering experiments at the Stanford Linear Accelerator Center discovered that protons and neutrons are made up of three quarks each. Quarks and leptons (electrons) are the elementary particles that make up observable matter in the universe, and both exist in three groups called the three generations of matter. Out of all the numbers available to the universe, why choose three?

It doesn't matter what atom of visible matter you examine, whether it's hydrogen, helium, or carbon. A stable existence requires that atoms have a numeric value of three or greater inside them. And if you think there's a law in nature that doesn't allow a hadron to accept more than three quarks during bonding, you haven't heard about the pentaquark. Every other hadron with fewer than three fermions, such as the meson, is unstable.

The cosmic atmosphere during the creation of matter (as seen in every atom) can be compared to a random Powerball result that draws the same three numbers out of a trillion every night. I would expect a universe without a fundamental intelligence or knowledge to make *at least* one mistake, wouldn't you?

The intelligence I speak of here should not be construed to mean one miraculously endowed upon the universe by a supernatural being. Rather, I would argue that mathematics is a process by which the universe applies intelligence.

For instance, a particle physicist may wonder why an inanimate universe would care about color neutrality as a reason for making a particle stable.

Should there even be a reason at all behind the decision made at random? I'm not saying a supernatural deity did it. My point is the acknowledgment of color neutrality by the universe could also be defined as intelligence or knowledge.

There's a difference between intelligence and an intelligent being. To understand this difference, we (humans) need to modify our common perception of what an intelligent being should look like. Just because life-forms are the only thing we recognize as intelligent beings doesn't mean a rock somewhere in the universe cannot be an intelligent being. We can't assume that every intelligent being must look biological in nature. If that's the case, it means we need biological cells to create our computers and artificial intelligence. So how do we know what has intelligence and what doesn't?

Intelligence is defined as the ability to acquire and apply knowledge and skills. "To acquire knowledge" means to be able to store information and to transfer and process information. Do you know that all the information in the universe is stored or transferred in a form of energy? So, consider this: What exactly is energy and an electric charge, and how did the universe come up with such a brilliant idea of a positive and negative? Do you know that the reason why you are referred to as intelligent being is because neurons in your brain receive, process, and send electrical impulses to the rest of your body via your nervous system?

Human beings are the most intelligent life-forms on Earth. How long will we continue to ask: "What is the real meaning of an electric charge? And what is the real meaning of energy"? If we don't have the boldness to answer this question once and for all, who will? Chimps? Aliens? Perhaps we can start by taking a small step toward a full understanding of intelligence and consciousness in the universe. Let me begin by redefining the meaning of intelligence and intelligent beings as they are observed in existence. (This definition is meant to apply throughout the universe.)

Intelligence begins when two or more bodies with an electric charge or any charge with similar attributes attract or interact by transferring, storing, and processing information (energy). The information they store or transfer is the intelligence. According to this definition, intelligence exists in anything that can store, transfer, and process information.

An *intelligent being* is any form of matter or energy that can independently use intelligence within its body to physically respond to its environment, acquire more information from its environment, and apply this information to solve a complex problem. In this sense, compare intelligence to software and compare an intelligent being to hardware. If you are a physicist, compare intelligence to bosons (the three force-carriers) and an intelligent being to fermions or hadrons. A human who is brain dead is an example of an intelligent being who lacks intelligence. What I'm implying here is that, the ability to solve a complex problem, or a mathematical precision in something is how we identify intelligence or intelligent involvement. For instance, in many part of the world, there are tall stone structures, such as mountains, that are shaped like a pyramid. But how did we know that intelligent beings built the great pyramids of Giza? The mathematical precision in the structures of the pyramids shows that intelligence skills were involved in the creation.

Mathematics, therefore, can be defined as the way by which intelligence is demonstrated in the universe. The fundamental constants that we observe in nature, and *Entanglement* are some of the ways by which the universe process intelligence. I explained how entanglement works in chapter eight.

From this point forward, when I use the word *God*, I mean mathematics (according to the definition above), knowledge, or power. I hope I was able to convince you a little bit about what it means to have intelligence. But just in case you are not fully convinced, I'll leave you with a thought-provoking question. Could intelligence as we know it exist on Earth if the elementary particles that make up the atom did not have an electric or color charge? In other words, if the electric or color charge inside the atoms that make up your body suddenly disappear, would you still have intelligence ability?

The elimination process

Let me tell you a story. Two illiterates were arguing whether the brain or the heart was responsible for memory. After what seemed like an endless argument, they finally came to a conclusion: if the removal of one of the organs caused a person to lose their memory, then it must be the organ responsible.

The next day, they went to a nearby hospital to see a doctor. At first, the doctor thought they had come for a medical checkup. But after listening to them go back and forth, he quickly realized he'd need more than a prescription to cure what was wrong with them.

There's an old saying that seeing is believing, so the doctor came up with an idea. The hospital had a public catalog of surgical videos for training purposes. He retrieved one of the videos from the cardiology catalog, *How to Implant an Artificial Heart in a Human,* and played it for them.

In the video, the surgical team removed a human heart and replaced it with a heart-shaped machine. Shortly after the surgery was over, the patient awoke, and the first question the nurse asked him had to do with memory: "What is your name?"

The patient replied, "Albert Ray the Third." All the nurses in the room applauded, and the video ended.

The doctor didn't know that the two illiterates had agreed on a conclusion before coming to the hospital—but to make sure they didn't go home and try to remove each other's brains, he invited them to the morgue.

Once there, he showed them one of the dead bodies. The doctor whispered gently to the two illiterates while they were staring at the corpse in awe: "This is what will happen if you *try* to remove your brain."

The title of this story is *The Elimination Process*. It's one of the methods used by detectives during investigations when they have multiple suspects. For example, during a homicide investigation, a suspect who will benefit the most from the victim's death is usually considered a strong suspect. With this in mind, let's treat the universe like a crime scene.

The elimination process as I will be using it in this book means that the correct order in which something in existence emerged can be arranged by categorizing existence into cause and effect, or Category one and Category two, respectively. This process would often reveal a suspect that cannot be excluded from the incident.

The universe is our victim and also our crime scene. Our suspects are the mysteries of how the universe works. Unlike Earth, where there are almost nine billion suspects, there are only three suspects in the universe: energy, matter, and space.

It may be difficult to recover evidence at a crime scene perpetrated by an intelligent being, as such a creature would know how to avoid leaving behind evidence. But if there is no secrecy involved in the existence of the universe, there should be tons of evidence lying around in every inch of the cosmos, waiting to be discovered. In most cases, finding these clues won't be easy. But do you know one thing I've learned about life? Anyone who hasn't faced tribulations hasn't endeavored to be great. It is as if the world is at war with the truth.

Most of the mysteries of our universe lie in physics. Our current physics is in a state I call *jigsaw puzzle*. We know the cosmic ingredients our visible universe is made of, but we just don't know how to put the pieces together properly in order to understand how the universe functions as a whole.

If you have the capacity to understand a simple process, such as how to kick a ball, you can understand any scientific process. There is no magic in physics. If you don't understand physics, either your teacher doesn't understand the problem at hand, or they don't know how to explain it to you in terms you can understand.

In case you don't have much knowledge about physics (none of us do), I'll try not to make this crime scene complicated.

The universe as a whole

Here is the universe in summary: space, energy, and matter. All other phenomena experienced in the universe are caused by interactions between these three. Space, energy, and matter are the fundamental foundation of existence that science is dedicated to understanding.

If we apply the elimination process to our universe, the question of how everything in the universe functions would simply become a question of stability. What does everything in the universe depend upon to function? Or what is it that, if removed, would make the universe cease to function? The three foundations of reality—space, energy, and matter—depend on force to function. If you compare the universe to a television, force is the electric current necessary to power it. Force is not an independent entity. This means it can only be applied through a form of matter or energy.

There has been much speculation on whether a theory of everything is possible. I believe there is more than one theory for everything, as there are many areas of study in science. However, if such a single theory did exist, it would be based on force.

From the bonding of particles to the orbits of planets, from the decay of particles to the adaptation of cells, force is why something happens in the universe. Thus, to understand how the universe works, we first need to understand how natural forces works, and how they apply to stabilize existence.

Stable Existence

Earlier in this book, I explained how the foundations of existence have three characteristics, and I pointed out why using the simple mathematics I referred to as the *multiplicative power*. Keep in mind that multiplicative power "starts" with three and is not limited to three. Every number greater than three also has multiplicative power.

Based on my observation, the multiplicative power is an inevitable characteristic that all *fundamental* components of existence must exhibit for it to be stable. Note that a fundamental question will have a fundamental answer. Thus, if it is presented to an intelligent life-form, such as a human, the answer will be as abstruse as the question itself.

Another way to understand the multiplicative power is to compare it to the ability of life-forms to reproduce. I can't give you a perfect theory of how life started on Earth; nonetheless, it seems to me that reproduction is a necessary characteristic that all living things must exhibit for life to continue.

I don't expect everyone to accept that the wisdom I extracted from the manipulations of abstract numbers is fundamental to existence. But show me something that's fundamental to existence, and I'll show you that it comes in three characteristics or more.

To ensure that stability continues in the things that are created, natural forces that are responsible for the functions of the universe are mediated through the relationship between "matter and space," or "energy and space."

Natural forces are often referred to as *forces of nature*. Forces of nature are the push and pull the universe uses to get things done. The force of nature does three things primarily: create, build, and change the building blocks of creation. We'll look into the relationship between matter and space to understand how force works in classical mechanics, and we'll look into the relationship between energy and space to understand how force works in quantum mechanics.

Matter and Space

If force is necessary to change the motion of a body, what force is responsible for the downward acceleration of an apple? This is the question that led Isaac Newton (1642–1726) to speculate on the force of gravity. He believed that for an apple to fall from a tree, an external force must attract it toward the center of the earth.

Later came the planetary question: What force is responsible for the motion of planets around the sun? Newton believed the same force of attraction that pulled the apple toward the earth was responsible for the orbit of planets. This idea was partially accepted until Albert Einstein, through his theory of general relativity, gave us an insight into why planets orbit without the need for an outside force.

Before we go into gravity, let's look at the three laws of motion by Isaac Newton.

First Law of Motion

Unless acted upon by an external force, a body at rest will remain at rest, and a body in motion will remain in motion.

Second Law of Motion

The force acting on a body is equal to the mass of the body multiplied by acceleration.

Third Law of Motion

For every action, there's an equal and opposite reaction.

Newton's Law of Universal Gravity

Every point mass in the universe attracts all others with a force that is directly proportional to the product of their masses and inversely proportional to the square root of the distance between them.

Nothing can be deleted or removed from the universe. Things only change from one form to another. Take weight, for example. A person who weighs two hundred pounds on Earth would have little or no weight in space. Where do you think the weight in that person went? I'm sure you know the universe is not Harry Houdini. Weight doesn't disappear and reappear like a coin in the hand of a magician.

Ever since I first learned about Newtonian gravity in high school, since the equations for calculating both are similar, I've been trying to understand the relationship between weight and force.

Whenever I think about why Isaac Newton applied the first law of motion to the falling of the apple, I wonder if Newton, who was sitting under the tree, would be surprised if he had looked up and saw the apple levitating or accelerating upward. I'm sure a levitating apple would scare anyone. That's because, naturally, anything that has mass when it is not suspended by a contact force is expected to fall.

Again, I thought about how the apple would behave in empty space. I imagined taking all matter out of the universe except for the apple. If I let go of the apple in empty space, would it fall, float, or levitate? Critical thinking such as this can take a toll on the mind, so occasionally, I'd go to the movies to relax my brain.

The movie showing one night was *Titanic*. I've seen this movie more than a hundred times, but for some reason, the film never gets old. One of the memorable parts is when the ship crashes against a floating iceberg. While everybody in the theater was engrossed in the tension-filled scene, what was going through my head were the laws of motion. How could an iceberg the size of a house float but a stone of any size sink? Why did a force that is said to act directly proportional to the product of mass make a 1 kg stone sink but couldn't sink a 1,000 kg chunk of ice? I ask this question because Newton's law of universal gravity doesn't state that gravity is directly proportional to the product of mass and density.

When the hammer and feather experiment was performed on Earth, the argument was that the hammer fell faster than the feather because it had *more mass* and air was present. Otherwise, in the absence of a medium, all bodies, regardless of mass, will fall at the same speed.

The proof of this came in the early 1970s when David Scott, one of the astronauts involved in the Apollo 15 mission, simultaneously dropped a hammer and feather on the moon. Both the hammer and the feather fell at the same speed and hit the moon's surface at the same time. This demonstration strongly suggested that relative density must play a role in explaining the theory of gravity. Simplified: I believe the reason why a stone sinks is also the same reason why the apple fell towards the earth.

During the days of Isaac Newton, when his theory of attraction had little or no rivals, people thought objects in space were floating (most people still do today). Thus, it makes sense to theorize that the earth causes objects to fall toward it. With this thought, it was easy for one to conclude that thrust force is needed for an object to escape the earth's pull. When

Albert Einstein got involved in the argument, we realized that with or without the earth, all objects in the universe are naturally falling. Therefore, with or without planets and stars around, thrust is required for a body with mass to accelerate upwards anywhere in the universe.

In space, the illusion of floating is created when two bodies fall relative to one another or when there's nothing in space to compare your motion to. For example, it's difficult for a human to see that everything in space is falling—until something like the earth appears below your feet. That's when it appears as if a force is pulling you toward the center of the earth. The truth is that you were falling even before the earth appeared below your feet; you just couldn't notice it.

As I have discovered, the biggest misconception in Newton's concept of gravity is the word *attraction*. It is obvious that a body with mass doesn't attract other bodies with mass like magnets attract one another. Therefore, it seems to me we are trying to condone the Newtonian force of attraction when we claim that gravity is "too weak." It's been more than 150 years now, yet the notion that two bodies with mass "attract" each other with a "force" has no perceptible *proof*.

The common layman's argument for gravity often goes like this: "If you don't believe in gravity, jump from a rooftop and see what happens to you." I understand this argument. I'm not arguing with *effect*. I'm arguing with the definition of the *cause*. Despite all the arguments I have put forward, Newton's equations of motion precisely describe effects but not the cause. Mathematical equations can describe *what* will happen, but not *how* it will happen.

Did you know that the same mathematical equation written to support the strong force would also work for a concept claiming that an invincible supernatural squid binds quarks together in a proton? On the other hand, the invincible squid theory can be disproven because we have *proof* that *gluons* (massless particles), not squid, bind quarks together in a proton. If someone makes a similar claim about Newtonian gravity, a mathematical equation alone will not be enough to save the theory of attraction, not to mention that falling is different than attracting.

If the force of attraction between two bodies with mass isn't similar to the exchange of gluons between quarks or the magnetic field between two magnets, the assumption is that the gravitational field between them is not propagating through space like gluons in a proton. If this is the case, unless we want to insert magical powers into science, Newtonian's force of attraction, according to its definition, is a reasonable misconception. Or maybe there's a yet-to-be-discovered quantum field carrying graviton particles.

Suppose a new quantum field called the graviton *is* discovered tomorrow. What are we hoping to prove with this field? That the graviton is the force of attraction between two fermions with mass and we can get rid of gluons? Is that true? That the graviton is the force of attraction between planets and stars, not the curvature of space, and we can get rid of general relativity? Is that true?

According to general relativity, gravity is explained by comparing free-falling celestial bodies to a bowling ball and the fabric of space to a trampoline. General relativity theorizes that, just as the mass of a bowling ball placed on the surface of a trampoline curves the fabric of the trampoline, a massive, rotating body would cause a curvature and distortion in space fabric, therefore causing other bodies with lesser mass to orbit around it. In this sense, space is described as a fabric that can be distorted or curved by the presence of a moving, massive body.

The gravity of general relativity, however, only explains the orbits of massive objects in space. The curvature of space-time did not explain why the apple fell, or what set mass particles in motion.

In this book, I try not to leave the explanation of gravity to the consideration of the reader. Instead, as an alternative to attraction, I theorize that in space, all bodies with mass are in a state of constant free-fall, or as I call it, *universal freedom*.

How this universal freedom works can only be understood properly if we know the actual problem in the theory of Newtonian gravity. The problem in Newton's idea of gravity simply lies in the definition, and the fact that our observable universe evolved from quantum processes to what we refer to as a *classic universe*. But we established most of the laws of physics with classical mechanics before we began a full study on quantum mechanics. The consequences of doing this are some of the problems we face today in physics.

Albert Einstein once observed that "Insanity is doing the same thing over and over again and expecting to achieve different results." If we hope to achieve a different result in the mechanics of physics, we need to amend or change some of the laws of physics to fit what we can explain in terms of a testable and demonstrable protocol.

As I have observed, mass and electric charge are fundamental properties of matter. Though electric charge isn't always present in some matter, a matter must have both of these properties in its definition. Because the definition of matter has to accommodate a new characteristic, the definition for mass must change as well. And because the definition of mass has changed, the definition of everything associated with mass must change accordingly. Therefore, the definition of magnetic, energy, matter, mass, density, momentum, and gravity would be amended as follows:

- *Magnetic* is the nature or ability of a body to push or pull another body that has a similar property configuration.
- *Energy* is an intangible, ubiquitous magnetic field responsible for the creation and functions of matter.
- *Matter* is anything that has mass or an electric charge and occupies space.
- *Mass* is the quantity of confined energy in a body.
- *Density* is the sum of mass and electric charge of any substance per unit of volume.
- *Momentum* is the sum of mass and electric charge of a body multiplied by its velocity.

Because matter and space are inseparable, according to a geometry known as space-time, natural acceleration of matter in space can be explained in terms of relative density. Regardless of what we believe, as long as we agree that space can bend, the fabric of space must have a density. However, such a density would be as difficult to explain as the nature of the fabric itself.

Before I proceed, let me state the unwritten rules that come with relative density: a body will accelerate in terms of momentum or change in position if it has more density than the volume it occupies. Accordingly:

- *Gravity* is the sum of density and momentum of matter or a quantum wave that allows them to move through the space fabric but in a direction allowed by constant velocity and the dimensions provided by space.

When using the definitions above, keep in mind that mass and energy are interchangeable, this means when dealing with a massless body, the entire energy of the body should be used to replace mass.

With this new definition of gravity, in classical mechanics, acceleration due to gravity, which I denote by italics "*G,*" has the same universal value of 6.674×10^{-11} Nm^2/kg^2 and would still be referred to as "acceleration due to gravity" or "relative density." In quantum mechanics, I will be using "G" to represent the speed of light "c^2". It will have the same value as 3×10.8m/s and will now be referred to as "acceleration due to gravity."

Why did I define energy as a magnetic field? As far as physics is concerned, there is no known energy field that doesn't contain a level of magnetic character, either as a particle or as a wave. Light wave, even though it has no magnetic moment, cannot exist without its magnetic field.

Gluons attract one another with their color charge, and this attractive ability could be defined as magnetic in nature. Even neutrinos, which have no electric charge, have a level of magnetic characteristics.

With these new definitions, you should begin to suspect a relationship between mass and electromagnetism. For instance, if someone tells you they observe a distortion in a section of space fabric but don't know what caused it, with your knowledge of general relativity, you are likely to tell them the presence and rotation of a massive body is the cause.

Likewise, if someone tells you they observed a flow of electrons in an electric field but don't know what caused it, with your knowledge of electromagnetism, you are likely to tell them the rotation of a magnetic matter is the cause.

But have you noticed that a body can have both mass and an electric charge, or have mass without an electric charge, but a body cannot have an electric charge without mass? With the elimination process, I conclude that an electric charge is a manifestation of mass. This means a body that contains an electric charge *must* have mass. And because mass could exist alone without electric charge, as seen in neutrinos, the electric charge of a body should be treated as a value that could be added to mass.

How gravity works

Classical mechanics

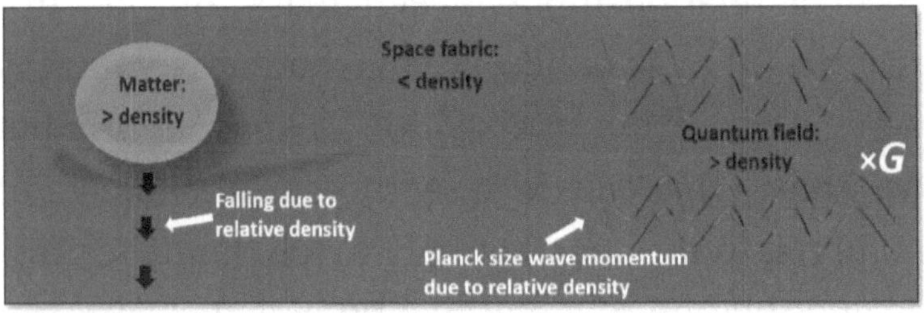

Figure 1.

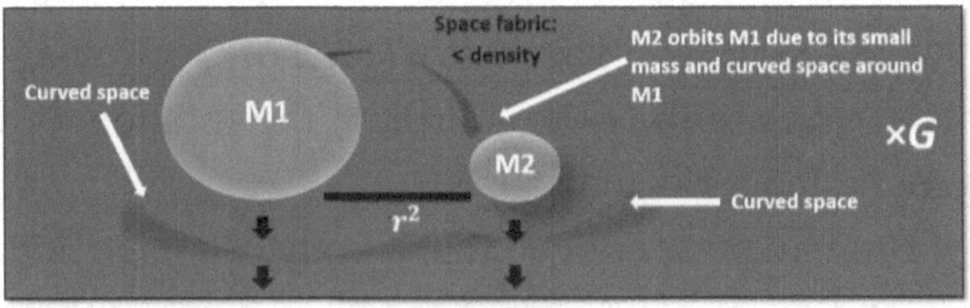

Figure 2.

Quantum mechanics

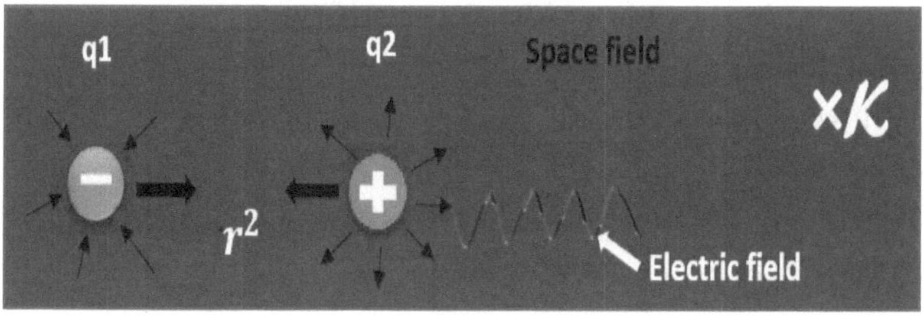

Figure 3.

Without a push or pull by an external influence, a body can only begin motion via relative density or its momentum. As seen in Figure 3 and figure 1 (quantum field), in a body where mass is small or equal to zero, motion is controlled by momentum multiplied by constant velocity, and in a body where mass is bigger, as seen in figure 1 (matter) and 2, motion is controlled by density multiplied by constant velocity.

The subscript "E" represents an electric charge, "M" represents mass, "V" represents volume or a field, "K" represents Coulomb constant, "*G*" and "G" represents constant velocity or acceleration due to relative density, and "*V*" represents velocity.

Density (d) = $M{+}_E / V$
Momentum (p) = $M{+}_E \times V$
Gravity (g) = $(M{+}_E / V + M{+}_E \times V)\, G$
$g = (d + p)\, G$
Where mass is small, $g_1 = p$, then p will be equal to $(E{+}_m \times V)$ G.
Where mass is bigger, $g_2 = d$, then d will remain as $(M{+}_E / V + p)$ G.

In a nutshell, gravity is not a force, rather it works according to relative density and momentum. For example, if you drop a stone in a swimming pool, the stone will sink (accelerate downward). I want you to compare the swimming pool to the fabric of space and compare the stone to massive bodies like planets or the apple.

Elementary particles (mass or massless), on the other hand, use momentum multiplied by constant velocity to accelerate through the fabric of space. Similar to how a human or fish uses undulation or locomotion to move in water, quantum waves use their momentum to propagate through space fabric.

Before you argue with this claim, imagine how a quantum body with no momentum would behave in space. Keep in mind that momentum and wave are inversely related. In theory, one can say, if there is no momentum, there is no wave and if there's no wave, there's no quantum momentum.

Acceleration due to gravity is based on a principle found in the second law of motion: "a body in motion will continue in motion unless acted upon by an external force." When a body begins motion in free space, it will automatically continue to move in that direction without the need of any external force.

In classical mechanics (g_2), when gravity sets a body in motion with a constant velocity, the body will automatically continue to accelerate with a constant velocity until acted upon by an external force. Likewise in quantum mechanics (g_1), when gravity sets a wave in motion with a constant velocity, the wave will automatically continue to accelerate with a constant velocity until acted upon by an external force.

Fourth law of motion

Isaac Newton believed that something is pushing or pulling on the apple, but in theory, the mass of the apple is the one pushing itself through free space. To better understand why this is so, I want you to examine the first law of motion again. If a body at rest needs an external force to begin motion, would an external force be needed to explain the motion of a body that can never be at or come to rest? The answer is no. Such a body would simply be a force carrier itself.

To take this a step further, we should probe the origin of matter in the universe. How are matter particles created? The two elementary particles—gluon and photon—where mass particles first originated (pair production) before they go through decay, are massless. In theory, this means they have been in motion since the moment they came into existence, and as far as we know, they have never been at rest.

This same principle applies to mass particles created from the reaction of massless particles. They came into existence in motion. Thus, a body (mass or massless) in free space and one that is not impeded by another body has a natural propensity to be in motion without the need for external forces.

This conclusion on gravity comes with a new law of motion, one which I believe would have been included in Newtonian mechanics if the full study of quantum mechanics had preceded the study of classical mechanics. *The fourth law of motion* states that in any field, a body in motion due to gravity needs an external influence to stop or change motion, but it would not require an external influence to be in motion.

Energy and Space

In the previous topic, we discussed how gravity works. But unlike the classic world, where mass is bigger and thus controls motion inside the space fabric, in the quantum world, mass is very tiny, and the magnetic ability of particles or electromagnetism, in most cases, controls motion in free space.

How electromagnetism works is not a problem. In fact, many physicists would agree it is one of the most understood concepts in physics. So instead of discussing how electromagnetism works, in this topic, I'll focus on explaining how the universe use the forces of nature to create and build matter.

Michael Faraday (1791–1867) and Paul Dirac (1902–1984) were some of the early contributors to the discovery of the electromagnetic field. As our science developed, we learned by observing a vacuum that a ubiquitous source must be responsible for various elementary particles in nature. All fundamental particles, including the ones we have yet to discover, emerged from a field inside the universe. The fields responsible for elementary particles are called *quantum fields*. The fabric of space itself, even though science cannot yet explain what it is made of, could be described as a field.

As I may be referring to space and quantum fields as if they are separate, I don't want you to look at them as two things that could be separated. Rather, think of both as an inseparable hybrid that make up the fabric of our cosmos.

Let me introduce you to the simple basics of quantum physics in my own unique way. Every nation has what is called currency. C*urrency* is "the system of money that a country uses."

Currently, the largest currency in the United States is the one-hundred-dollar bill. But if you study the bill—let's call this study the "universe"—you will discover that it can be divided into two fifty-dollar bills. Two fifty-dollar bills can be divided into five twenty-dollar bills. A twenty-dollar bill can be divided into two ten-dollar bills. A ten-dollar bill can be divided into two five-dollar bills. A five-dollar bill can be divided into five one-dollar bills. A one-dollar bill can be divided into two half-dollar coins. A half-dollar coin can be divided into five dimes. One dime can be divided into two nickels. One nickel can be divided into five pennies.

If you study the penny—let's call this study "quantum physics"—you will notice that it cannot be divided further. In the United States, a penny is the indivisible coin that makes up every other currency. Compare a fundamental particle to a penny.

Quantum physics is the study of the indivisible particles that make up everything inside the universe. Anytime you hear the words *fermion, boson, fundamental particle,* or *elementary particle,* compare them to a penny. Elementary particles are often referred to as *wave particles*. This is because they sometimes behave like a wave, and sporadically can behave like a particle. In addition, an elementary particle can have a positive charge, negative charge, or no charge.

All observable elementary particles are referred to as *quarks, leptons,* or *bosons.* Quarks and leptons are referred to as *fermions*. Fermions come into existence with an opposite of the fermion known as *antiparticle* or *anti-fermion.* Bosons are the force-carrying particles that allow quarks and leptons to interact. Bosons don't have antiparticles.

The way to differentiate between an American and British penny is by looking at the color and inscriptions on the coin. The way to differentiate one elementary particle from another is to look at the color, charge, spin, or mass.

If you have a lot of pennies but need a nickel to make a payment, you need to add five pennies together. In quantum physics, adding pennies can be compared to the bonding of fermions. If you need to break a nickel into five pennies, you divide the nickel by five. This process of breaking down a nickel into pennies can be compared to decay in quantum physics.

In quantum physics, the plus sign (+) that helps convert five pennies into a nickel is known as the strong force. The division sign (÷) that helps break down a nickel into five pennies is known as the weak force. Just as the penny is produced in the United States, elementary particles are produced from quantum fields.

The best way to visualize a quantum field and how it produces wave particles is to compare it to a game called Whac-A-Mole. The quantum field is like the floor in this game, and wave particles are like a mole that keeps popping out of a hole in the floor.

During the interactions of fermions, force-carrying particles are exchanged. W and Z bosons are the agents of the weak force, while gluons and photons are the agents of the strong force and electromagnetic force.

In quantum physics, physicists use a method called the Feynman diagram, invented by American physicist Richard Feynman (1918–1988), to visually illustrate the interaction between fermions, hadrons, and bosons. *Hadrons* are mass particles made up of two or more fermions.

The Higgs particle is like the commander in chief of the armed forces. I like to refer to it as the "big for nothing else" particle. This unstable particle has the biggest mass, which, in turn, means its quantum field is responsible for the majority of the mass of all other elementary particles.

Neutrino particles are like cryptocurrencies. Even though we know what they are and how they are produced, at the time of this writing, they are not legal tender. We use cryptocurrencies to trade via the internet. In quantum physics, neutrinos are known to have no charge and only interact via the weak force.

Like-charge particles repel, while opposite-charge particles attract. The strong force bonds quarks together, while the electromagnetic force only exists where there are electrically charged particle interactions. The weak force is responsible for the decay of particles into particles of less mass.

These are the summarized basics of quantum physics, and ones that you will need to understand some of the explanations in the next chapter.

Chapter 5

The Multiplication Theory

I invented the *multiplication theory* as a means to understand the fundamental principles of the universe. The multiplication theory is a basic concept that uses the laws of existence, coupled with mathematics, quantum, and classical mechanics, to explain the fundamental principles of existence. The multiplication theory is discussed at length later in this chapter.

As it has been observed in nature, matter and energy are the same property in different forms. Though the entire universe may be too complex to comprehend by our tiny minds, we understand the mechanics behind the transformation of matter into energy and vice versa.

We have probed the atom to the indivisible properties that it is made up of, and as far as we can tell, there's no matter we know of in the universe that is not made up of energy. Thus, it is fair to come to a conclusion. As fermions and bosons cannot be divided further, I refer to them as *creating properties*. In this sense, fermions are like bricks, and the bosons that bind them together are like mortar. What two or more fermions or particles become after they are bound together by a boson, or by simply exchanging a fermion, is a *creature* or a *created thing*. Any high oscillating boson field is equal to a creator, and the reactions of a gauge boson, known as pair producing, are equal to the first creatures. Accordingly, a creator can be said to be equal to the creature it creates but different in appearance.

The merger (mathematics)

It's important to remember that we use mathematics as a language to understand reality. Thus, arithmetic signs, shapes, and operations of mathematics don't make sense if they cannot be translated into something real.

In fact, the languages of mathematics are used to understand the value, shape, or quantity of a tangible or intangible thing in existence. In this sense, mathematics is like the definition of something. You can also think of it as a shadow. When you see a shadow, you know it indicates the presence of matter. The same is true with mathematics. It is a language used to describe the shape, size, value, or quantity of something tangible or intangible.

In the multiplication theory, we will be dealing with the translation of abstract numbers into something real. The best way to interpret abstract numbers is to match them with something tangible and intangible that perfectly and universally describes quantity and value.

Based on our current understanding of the universe, the shape, value, size, and quantity of energy, matter, or space are what mathematics is used to understand. And out of the three foundations of reality, only energy and matter are quantized. Space is mostly represented by vector.

Hence, since this theory requires that we replace numbers and operations with a form of existence, I will be replacing mathematics with quantized matter and energy.

In our observable universe, and as far as we know, this is the order in which the tangible and intangible representation of numbers exists in reality: $\underline{0}$, $\tilde{O}$, $\bar{o}$, -½, ½, 1, 2, 3 . . . etc. These numeric values describe fermions, bosons, and hadrons. In reality, these numbers are so small that the best way to describe fermions and bosons is "second to nothing." If you try to go smaller than a fundamental particle, you will arrive at nothing (fields). Those three non-numeric values—the 0s—represent space fields, fermion fields, and boson fields.

These fields are not absolute zero, as you may have thought. When they are not under the influence of a disturbance, such as force, their intangible representation in mathematics would be 0.00—something that has value, or as understood in physics, Planck time.

Since we don't know the exact mathematical value below zero, let's just call them *zeros*. The lines above and under the three zeros represent Planck time. Thus, when a quantum field is not under the influence of a noticeable force, the value of these fields in calculations should be replaced with the Planck time constant.

The value of the smallest observable matter in existence is mathematically represented by ½, or half. This lets you know that 1 is made up of two values; that's why the arithmetic sign of half is denoted by 1 divided by 2. In reality, -½ and ½ represent anti-fermion and fermion wave particles. The 1 represents a gauge boson wave particle. The 2 represents a meson particle, and the 3 represents a baryon particle.

The 1, therefore, is a description of -½ and ½. The 2 is a description of any two ½'s bound together by 1. The 3 is a description of any three ½'s bound together by 1, and so on and so forth. In this sense, 1 serves as a force-mediating particle. Unlike a fermion field, which produces -½ and -½ wave particles simultaneously, a force field produces 1 gauge boson.

Why did the multiplication theory use quantum mechanics to replace mathematics and not classical mechanics? Multiplication theory is a concept that is created to understand the universe; therefore, it requires a universal representative for mathematics.

For example, if 1 is represented by an orange, the description will hold until the orange is divided among ten people. Thus, to avoid an infinite number of 1s, the first law of existence requires that we represent the smallest numeric values with something in existence that cannot be divided further and that we represent 0 with something that describes the origin of 1. In reality, -½ and ½ are reactions of 1, not something that 1 split into.

In case you are wondering why I stopped at ½, a boson (1) only produces two particles, known as *pair production*. These two particles are also the smallest things known to humans. In reality, there is no such thing as ⅓, ¼, and so on and so forth, of a gauge boson.

The mathematical representation of nothing is zero (0). This needs to be clarified again before we continue. When someone uses the word *nothing* to describe space, what the mind often thinks about is the absence of everything tangible and intangible. As I have discovered, *nothing*, according to how it's viewed, doesn't exist in the universe. It is impossible to prove that nothing exists anywhere in the universe. Accordingly, nothing cannot be proven when it is defined as the absence of something. It can only be proven when it is defined as the absence of something specific in an area of space. The absence of that specific thing in an area of space can then be referred to as nothing.

For instance, just because you've emptied your pocket at the airport before going through the metal detector doesn't mean there's nothing left in it. Your pocket might not contain your wallet and coins, but it does have air. You can remove all the air in your pocket if you are not satisfied, but you will need divine intervention to remove quantum fields anywhere in the universe.

It is easy to think about or use the word *nothing*, but in reality, even in the so-called vacuum of space, no one can demonstrate a place, area, or point that doesn't contain anything tangible or intangible.

Quantum fields are indelible properties of space. Therefore, *nothing* or *zero*, as it relates to existence, can be defined as an indelible, palpable, intangible field that allows the occupation of matter. These fields are divided into three: space fields, force fields, and fermion fields. (Force fields are the same as boson fields).

By exchanging field momentums, two fermion fields can cause a boson field to become energized and oscillate rapidly. In exchange, the energized boson field causes a fermion field to oscillate more quickly.

The observable oscillations of both fields are wave particles.

Translated into the language of mathematics, a fermion field momentum plus another fermion field momentum is equal to a high oscillation in boson field, or $\bar{o} + \bar{o} = \Delta\underline{0}$.

The change in field here simply means that fermion field interactions cause a boson field to become energized and oscillate rapidly. In return, the reaction of an energized gauge boson caused a fermion field to oscillate rapidly, $\Delta\underline{0} \times \bar{o} = \Delta\bar{o}^2$.

Simplified, if you bang the nature of nothingness hard enough, something noticeable will come out of it. The kind of action necessary to cause an intangible fermion field to experience high oscillation is what I am representing as multiplication in this book.

In this theory, we will look at numeric values as wave particles and arithmetic operations as the actions of bosons. Any value that starts with zero (0) is a field. A single fermion that cannot be divided further is represented as half (½). Two fermions bonded together are represented as two (2). Three fermions bonded together are represented as three (3). Any boson particle is represented as one (1).

Think of the hybrid of mathematics and quantum mechanics this way. Numbers are mostly used in mathematics to understand the value or quantity of something observable. Arithmetic operations are what humans use to manipulate numbers. Fermions are the building blocks of observable matter. Bosons are what the universe uses to manipulate fermions.

Hadfer Table

"Hadfer" is a neologism derived from combining the abbreviation of *had*rons and *fer*mions. The Hadfer table uses numbers (as they exist in reality) to test the strength of the two operations of creation (×, +).

The Hadfer table illustrated later in this chapter shows how numeric values are used to test the strength of addition and multiplication as discussed in chapter two. Look at the results of numbers in the Hadfer table as a hint of what to expect from certain types of bosons in reality, not the exact result you would derive in reality.

Let me use something that you might be familiar with to explain this. Have you ever had a dream that came true? You probably noticed that what happened later in reality wasn't the same as how it happened in the dream.

For instance, maybe in your dream, you won $50 million. Because of that dream, you played the lottery the next day on your way to work and won $1,000. Even though you didn't win $50 million, the dream gave you a hint of what would happen in reality if you played the lottery. Such is mathematics. It's like a dream. Calculus will show you a *hint* of what to expect in reality.

When you see the numbers in parentheses in the Hadfer table, keep in mind that 1 is a force-carrying particle, while 2, 3, etc., are *descriptions* of how many ½'s are in a particle.

Since 1 isn't the smallest numeric value and, in reality, also represents multiplicative force, 1 will not be used to test the strength of a force in the Hadfer table. The reason for this is similar to how it is allowed in mathematics to multiply addition and subtraction signs with each other and by each other. But there's no demonstration I know of where the multiplication sign is manipulated by another arithmetic operation or by itself.

I only focus on addition and multiplication in the Hadfer table because they are the only operations that can increase a numeric value. In reality, this means the act of multiplying and adding are the only actions that can be used to create something.

Mass and electric charge

In quantum mechanics, the factor that determines if a fermion is greater or lesser is mass. Mass, as I explained earlier, is related to an electric charge. This is why the mass of fermions is measured in electron volts. Electron volts are like punches. The more punch you pack, the harder you will be able to bang a fermion field to produce a fermion.

In the Hadfer table, I want you to relate the results of multiplication to the strength or mass of a particle that a multiplicative force would produce during a high-energy interaction or collision.

When a body mediating a force such as bosons also carries an electric charge, its reactions (fermions) will not be accurate. Just as mass can push something, an electric charge can push or pull something as well. However, in the quantum world where matter is very tiny, the effect of an electric charge is stronger than mass.

Compare the electrical charge of a boson to this scenario. Anything with mass has the ability to apply force, including you. So, if you stand motionless on a scale and then switch it on, the number that appears when it comes on will be equal to your weight. But suppose the scale is turned on and you jump on it, or you keep moving while you are standing on it. In either of these cases, the number on the scale will not equal your actual weight. You can compare your motion while standing on a scale to the electrical charge of a boson.

It's not a problem for fermions to retain an electrical charge, but when the actions that produce them possess an electric charge, it will affect the accuracy of the reactions. An electric charge, therefore, is a disturbance when possessed by a boson.

Naturally, matter can exert force when it comes in contact with other matter or space. However, if a form of matter possesses an electrical charge, it can exert more force than just its contact force. What this means in a quantum reaction is that an electrical charge affects the reactions of a boson.

An electrical charge contributes a lot in the quantum world where mass is barely noticeable, but when it is possessed by a force mediator, it is a disturbance. Hence, provided no internal force disturbs an action (boson), for every action, there are equal and opposite reactions. Accordingly, as the third law of motion is true in classical mechanics, in quantum mechanics, we should find an undisturbed force carrier producing equal and opposite reactions.

<u>Origin of Particles</u>

Boson 1, fermions ½, mesons $\underline{2}$, baryons 3.

The line under "$\underline{2}$" describes it as bosonic.

<u>Quantum Field Interactions (QFI)</u>

P = momentum

Boson-field momentum = $\underline{\mathit{0}}$ (p)

Fermion-field momentum = $\bar{o}$ (p)

Classical mechanics, $G = 6.67\times10^{-11}\ \mathrm{Nm^2/kg^2}$

Quantum mechanics, $\mathrm{G} = 3\times10.8\ \mathrm{m/s}\ (c^2)$

$E = mc^2$

$\bar{o} + \bar{o} = \Delta\underline{\mathit{0}}$ (momentum exchange)

$\Delta\underline{\mathit{0}} \times \bar{o} = \Delta\bar{o}c^2$

$\Delta\bar{o}c^2 = 1c^2$

$\bar{o} + \bar{o} = \Delta E$

$1c^2 = 2\psi c^2$ (pair production)

ΔE = two fermion field exchange momentums

h = Planck constant, f = frequency of a boson

$E = hf$, $E = mc^2$, $F = ma$

QFI equation, $\Delta E/h \; \alpha \, f = 2\psi c^2$

"G" is the same as c^2. "E" in $E = hf$ is equal to $\bar{o} + \bar{o}$ or ΔE, which is directly proportional to *f*. "*f*" is the frequency of a gauge boson, which when it reacts is equal to mc^2 or $2\psi c^2$ (two mass particles multiplied by the speed of light.)

Hadfer Table

(½+-½) = $\underline{2}$	(½×-½) = 1
($\underline{2}$+$\underline{2}$) = 4	($\underline{2}$×$\underline{2}$) = 4
(3+3) = 6	(3×3) = 9

Just as an arithmetic operation such as the addition sign (+) can be multiplied by itself in mathematics, the power of addition in a boson can be tested. However, because addition also has a multiplicative ability (as seen in the Hadfer table when both addition and multiplication produce the same result), multiplication (×) cannot be multiplied using the multiplication sign.

In other words, we cannot collide a fermion or hadron to test the power of addition in a boson. If we do, we will end up performing an unintentional multiplicative test.

In reality, the addition test above is equivalent to allowing two fermions or hadrons to gently interact so the strong force (+) can take over. Then we can observe the reactions of the bosons mediating the force between them and compare it to multiplication.

To test the multiplicative strength of a boson, you need to collide two fermions, which could produce the boson, or two hadrons carrying the boson. The multiplicative reactions of a boson, specifically, the mass of the fermion produced, will be greater depending on the multiplicative power or the generation of the fermion carrying or producing the boson.

Bosons (multiplicative forces) are packs or punches of energy. Thus, they can exhibit the act of multiplication if two of them interact. However, the punch of a boson increases depending on how many fermions are contained in the hadron that is carrying it. During multiplication, the more fermions a particle contains, the more powerful the reaction a multiplicative force within it will produce. For instance, the reaction that two gluons will produce if they casually interact will be different from the reaction they will produce in a proton-proton collision.

If you imagine that the results we derived from adding and multiplying the numbers in the Hadfer table are a dream, when we replace arithmetic operations with bosons, we should expect to see a similar reaction between the boson that represents the addition and multiplication operations. We should expect to find two opposite fermions bound together to be unstable. We should expect the multiplication of two opposite fermions to produce a type of boson. We should expect a greater reaction from the multiplication of two particles containing three or more fermions.

Since no symbol represents the three generations of matter, I denote all fermions and anti-fermions by wave function ψ. Although I used "ψ" in the field equation to represent fermions, in calculations, they should be replaced with the wavelength or amplitude of a wave particle.

Bosons and Electrical Charge (*EC*)

"1" in the Hadfer table means an undisturbed gauge boson.

$EC \leq 0$: photon = equal and opposite reactions

$EC \leq 0$: gluon = equal and opposite reactions

$EC \leq 0$: Z^0 = equal and opposite reactions

$EC > 0$: $\pm$ W = non-equal and opposite reactions

Equal and opposite reactions mean pair production (ψ). Non-paired production ($\boldsymbol{\psi}$) means non-equal and opposite reactions.

F = force carriers or gauge bosons

$|F = \gamma|$ $|F = \pm W|$ $|F = Z^0|$ $|\psi, \boldsymbol{\psi} = a|$

$|_1 \pm w = 2\boldsymbol{\psi}|$ $|_1\gamma = 2\psi|$ $|_1 Z^0 = 2\psi|$

F = ma. F = force, m = mass, a = acceleration

MTE = multiplication theory equation

Classical mechanics, a = acceleration

Quantum mechanics, ψ = wavelength or amplitude

$a = \psi, \boldsymbol{\psi}$

$2a$ = equal and opposite reaction in classical mechanics

2ψ = pair production

2ψ = non-pair production
MTE 1, $F = [(m_1 \times m_1) / (m_1 + m_1)]\ 2a$
Massless particle $E_1 = E_K$
Kinetic energy $E_K = \frac{1}{2}mv^2$
Constant speed K_S = constant K_E
Constant force $K_F = K_E$
MTE 2, $K_F = [(E_1 \times E_1) / (E_1 + E_1)]\ 2$

The merger (Classical mechanics)

The most important thing to understand before you read the multiplication theory is motion. A wave is a type of motion. Every motion, whether in classical mechanics, quantum mechanics, or alien or religious mechanics (if they exist), would obey at least one law of motion.

To understand why quantum fields oscillate faster, why wave functions collapse, and why particles of matter are produced in pairs, we must go back to the backbone of physics, to the laws of motion, which are often ignored in quantum mechanics.

The bridge between classical and quantum mechanics lies in the multiplication theory, which focuses on merging the two. To understand how this is done, you have to remember that quantum waves are a type of motion and, as such, the laws of motion apply to them.

It is reasonable to apply a known cause that produces an effect to a similar effect that one sees but may not understand the cause. For instance, a moving golf ball suddenly comes to rest. Even if you can't prove the force that brought the ball to rest, you can confidently say the motion of the ball was affected by an external force and that is why it suddenly came to a stop. Why can't we say the same when we observe the collapse of a wave function?

I understand that some observations in the universe are unpredictable with our current physics. And as I often say, no matter how good your theory about the constantly changing universe may look now, it will eventually find a way to embarrass you. The good part is that the universe doesn't know how to break its fundamental laws. That's why we are here.

On the other hand, I also understand that within the same reality, the principles that govern a fundamental law may adapt to changes, but the principle itself cannot change. For example, it is a fundamental law of matter that for a body having mass to travel from one point to another, the body has to undergo motion. Let's say it took me (a human) ninety-nine hours to run from Los Angeles to New York City.

The next day, I found a magical power that turned me into an electron. When I return to Los Angeles from New York City the following day, the trip only takes me less than a second because I'm an electron.

In this example, notice that motion had to adapt to the change in mass in terms of speed and time, but size did not change the fundamental law of matter.

I mentioned earlier that even if you can't prove the force, you can still confidently say the motion of the golf ball was affected by an external force, and that's why it suddenly came to rest. Isaac Newton did this when he formulated the force of gravity, and around this time, he invented calculus. Newton cannot explain how gravity works, and even today, no one can prove the existence of attraction in gravity. Yet, the math works in explaining effect. If we approached classical mechanics this way, there's no reason not to apply the laws of motion to quantum mechanics. In conjunction with the new definitions I provided earlier in chapter three, accordingly:

- The first law of motion applies to quantum fields in that at a constant time, the change of a quantum body with conserved momentum will remain as such, and a quantum wave in motion will remain in motion unless acted upon by external forces.
- The second law of motion applies to quantum fields in that when a quantum body is not under the influence of external action, the change experienced by the body is equal to the sum of its density and momentum. For a quantum body under the influence of external action, force is equal to probability of the change in the act of creations multiplied by double acceleration.
- The third law of motion applies to quantum fields in that for every undisturbed gauge boson, there shall be equal but opposite reactions, and for every disturbed gauge boson, there shall be non-equal and opposite reactions.

By applying the laws of motion to quantum fields, gauge bosons are described as actions, and every fermion and anti-fermion in the standard model are described as equal and opposite reactions of gauge bosons.

Paul Dirac (1902–1984), an English theoretical physicist, was the first to predict antimatter through abstract mathematical equations that he derived from merging special relativity with quantum mechanics. In this book, I only explain *why* it exists with the multiplication theory.

In the multiplication theory, "multiplicative cause" is the same as "multiplicative force," and multiplicative effects are the same as multiplicative *reactions.*

Multiplication theory

As mentioned, the multiplication theory is a basic concept that uses the laws of existence coupled with mathematics, quantum, and classical mechanics to explain the fundamental principles of existence.

It states that multiplication is a fundamental act of creation that uses multiplicative cause to produce multiplicative effects. Based on mathematical demonstration, it shows that a multiplicative cause that undergoes multiplication through a multiplicative power has the ability to bring something massive out of the intangible nature of nothingness; therefore, multiplication cannot be excluded from the cause behind the first tangible mass creature in existence.

In this theory, non-numeric values or nothingness (0) represent fields, numeric values represent fermion and boson wave particles, and arithmetic operations represent bosons and their actions.

This theory adds that, according to the third law of motion, because force is an action that produces an equal and opposite reaction, the first tangible multiplicative effects would simultaneously come into existence in two pairs, each representing the equal and opposite reaction of force, and during their lifetimes as mass particles, they would evolve either through multiplication or addition, which are the two acts of creation.

In this theory, the addition and multiplication operations represent the only two actions of creation that a multiplicative force can possibly exhibit in reality. And because both actions have equal reactions which, when combined together, are incompatible, any particle that represents the numerical value that describes the combination of two equal but opposite reactions will be unstable.

The idea is that the universe operates on this mathematical principle and, as a result, particles will not be stable unless made up of multiplicative reaction, force, or power.

In this theory, -½ and ½ (fermions and anti-fermions) are multiplicative reactions, where 1s (gauge bosons) are multiplicative forces, 3s (baryons) or any numeric value higher than that are multiplicative powers, and 0 (fields) are the nature of nothingness that fermions and bosons occupy and originate from.

Any gauge boson that binds two fermions together exhibits addition behavior, and any boson that produces pair production during a reaction exhibits multiplication behavior.

Consequently, the act of adding occurs when two or more fermions bind together, and the act of multiplying occurs when a high- or low-energized boson starts a chain reaction that begins with pair production. The higher the multiplicative power used to energize the boson, the greater the mass of the multiplicative reactions a multiplicative force will produce during multiplication.

Anti-fermion (antimatter)

Particles of matter come into existence simultaneously as separate, single particles called fermions and anti-fermions (pair production). The antimatter of hadrons formed in the same way the hadrons formed. For example, the same way two up and one down quark combine to form a proton, two anti-up and one anti-down quark combine to form an anti-proton.

Think of the existence of anti-fermion (i.e., positron) as rocket exhaust and the actual fermion (i.e., electron) as the rocket. The exhaust explosion that pushes a rocket up is equal to the mass of the rocket but opposite the direction in which the rocket is traveling. This phenomenon, which is stated in the third law of motion, is also why pair and non-pair production happens in a quantum field.

How multiplication works in quantum fields

The most important thing in existence is force. Without force, space would be void of something tangible. Force creates motion, and motion creates time. Time is how we make sense of space and all the changes happening around us. You and I, and all the matter you see today in the universe, began with some natural forces. Ask any physicist what these forces are and they will say: strong force, electromagnetic force, weak force, and gravity. Trust me, they can explain how these forces work.

Except Newtonian gravity, they can even show you gauge bosons, particles that mediate these forces, and explain how each of them interacts with particles of matter called fermions. If you happen to be politely stubborn, like myself, you can take this one step further and ask: What force instigated the oscillation of gauge boson and fermion waves? Trust me, any physicist will tell you that, according to quantum theory, bosons and fermions are oscillations in different quantum fields spread across every inch of space in the universe.

Now, at this point, I can only predict what a person would say if you asked where quantum fields came from. My best guess is the big bang theory.

This is where the multiplication theory comes in. The multiplication theory says that you don't need the act of creation to explain the foundation of existence. Because the nature of nothingness just happens to be occupation (space), cause (boson field) and effect (fermion fields), which are fundamental bedrocks that are needed to begin creation, the theory says if you apply the first law of existence—classical mechanics and mathematics to quantum mechanics—you can begin with space and quantum fields and explain the existence of the universe as it is today without the need to explain how the universe (space-time) itself began.

The theory says one of the natures of nothingness (fermion fields) produces gauge boson particles by adding tiny momentums, and in return, gauge boson particles produce particles of matter with fermion fields through the act of multiplication, and these particles would continue to evolve into something more observable by way of addition and multiplication.

But how does multiplication work in quantum fields? In mathematics, it's easy to say 7+7 = 14, until you ask what binds 7 and 7 together to form 14.

If you try to understand this question with realistic demonstration, the quest will lead you into quantum mechanics. You will find that the plus sign (+) used to combine numbers has a physical representation called the *strong force* and it is mediated by a massless quantized energy called gauge boson. All gauge boson particles, which are often carried or emitted by fermions and hadrons, are what the multiplication theory above refers to as multiplicative cause (aka multiplicative force).

Since the act of multiplication begins with any action that could energize a boson up to the point where it reacts, any gauge boson with the ability to produce pairs of equal fermions has the ability to multiply. However, the chain reaction of multiplicative force will be higher or lower depending on whether the chain reaction starts from the first, second, or third generation of matter. The higher the multiplicative power, the greater the mass of the multiplicative reactions a multiplicative force will produce during multiplication.

In this sense, I want you to think about multiplication as an action. To better understand how this action works, think of a multiplicative force as a bullet, and think of multiplicative power as gunpowder. The more gunpowder a bullet shell case packs, the more damage the bullet will cause. But unlike a bullet, multiplicative forces don't get their power from explosion; rather, they become energized by the kinetic energy they generate from being passed around at the same speed and time.

When an object moves, it generates kinetic energy and the temperature increases. The faster the object moves, the higher its kinetic energy and temperature. For example, if you shake a basketball repeatedly, after a while, the temperature of the air inside the ball will increase to, let's say, 2°F. With one hand, if you bounce the basketball continuously against the floor at the same speed, the temperature will increase to, let's say, 4°F. With two hands, if you bounce the basketball continuously against the floor at the same speed, the temperature of the ball will increase to, let's say, 6°F. The idea is that the more contact force you apply to the ball at the same speed, the more energized the air inside the basketball will become.

Now let's bring the analogies above into quantum mechanics. The energy level of a photon or gluon (multiplicative force) emitted by one electron depends on the momentum of one fermion. If a gluon inside a meson has been passed around with equal speed and time by two quarks, it would make contact with two momentum and thus would be more energized than an ordinary gluon emitted by a single quark or a single photon emitted by one electron.

Likewise, if a gluon inside a proton has been passed around by three quarks at equal speed and time, it would make contact with three momentum and thus would be more energized than a gluon inside a meson.

When we look inside a proton, the collision of two gluons is constantly undergoing multiplication by producing quarks-antiquarks from the first and second generation. On the other hand, inside a meson, if we could get a picture of the interior before it decays, the collision of gluons would produce quarks-antiquarks in the first generation.

When man or nature take multiplication one step further by colliding two fermions or hadrons at the speed of light (c^2), a multiplicative force emitted or carried by any of these particles will become more energized in addition to its current kinetic energy and as a result, multiplication will occur and continue through a sequence of chain reaction via the weak force. This chain reaction, where in most cases both pair and non-pair production are produced from different energized bosons, is how multiplication happens in quantum fields.

A boson emitted by the collision of two fermions at the speed of light can be mathematically expressed as $\Delta\psi^1(c^2)\ /h = m_1$. A boson emitted by the collision of two mesons at the speed of light can be mathematically expressed as $\Delta\psi^2(c^2)\ /h = m_2$. A boson emitted by the collision of two baryons can be mathematically expressed as $\Delta\psi^3(c^2)\ /h = m_3$.

The kinetic energy of a massless gauge boson is denoted as $\Delta\psi$ (change in kinetic energy of frequency). $\Delta\psi$ is multiplied by the numbers of fermion exchanging the boson in a hadron or a single fermion emitting it. A mass from the first, second, and third generation of matter are denoted by m_1, m_2, and m_3. These generations are arranged from the smallest mass to the highest mass known to date.

To a mathematician who only deals with abstract aspects of reality, the act of multiplying can only be done by recounting the available numbers multiple times. The universe, however, doesn't have this abstract luxury of recounting real things to produce more. The universe is real, so must its result when it performs multiplication.

In the biology aspect of existence, if you place a single bacterium on a slice of bread and leave it in a kitchen at room temperature, in less than two hours, a single bacterium will have become thousands. But unlike bacterium, which multiply by splitting its nucleus, in the reality of quantum fields, multiplication occurs when the reaction of an energized gauge boson creates pair production. The chain of multiplication would continue via the weak force depending on whether the pairs of fermions belong to the third or second generation of matter.

The relationship between gauge bosons and fermions are direct, such that one can trace every fermion and anti-fermion back to every gauge boson, and vice versa. Bosons are causes, while fermions are effects. Or, one could say, bosons are actions, while fermions are reactions.

For instance, because photons and Z bosons produce equal and opposite reactions. If the reactions produced by any of them are reversed, photons or Z bosons can be produced from the annihilation of any first- and second-generation fermions with their corresponding anti-fermions.

The same applies to the W boson when non-equal and opposite fermions collide. The gluon, however, since it is known to only react in quark fields, can only be produced during quark and corresponding anti-quark annihilation and can bring into existence quarks from the third generation when a multiplicative power is involved.

If a multiplicative power greater than three, such as exotic hadrons, are multiplied (collided at the speed of light), a multiplicative force within it should be able to bring into existence a fermion from the fourth generation ($\Delta\psi^4(c^2)/h = m_4$)—*if* the fourth generation exists in the universe. This direct correlation between fermions and bosons, as explained by the multiplication theory, is what I call *direct symmetry*.

The QFI equation, $\Delta E/h \; \alpha \; f = 2\psi c^2$, will not include a second particle (antiparticle) when applied to a force-carrying particle. This is because bosons are the cause of pair or non-pair production; therefore, they cannot have separate antiparticles. In other words, a gauge boson will be its own antiparticle.

As mentioned, an equal reaction needs an opposite reaction to exist, and vice versa. For example, you need to pull a bow backward for the arrow to shoot forward. But before you pull the bow backward, you don't have to pull something else backward. Otherwise, you would need an infinite number of backward pulls to shoot the arrow forward. Such are the actions we refer to as gauge bosons.

When a fermion like the electron emits a photon, or during decay when a fermion emits a weak force particle, the fermion in each case continues to exist for a fraction of time enough to allow the boson to exist. This creates a balance action. Balance action, as I use it here, means quantum field requires two things to be in existence for one to exist. Which in turn means one would have the ability to produce two things to prevent this balance from being violated.

If one particle needs to create or emit another, that particle must stay in existence for a fraction of time enough for the other particle to exist. Otherwise, the particle that is created would immediately become two particles if the emitting or creating particle cease to exist immediately.

Another way a boson is created is by two fermion exchanging momentum. If the boson is energized enough, its reaction will create pair or non-pair production. Thus, because a boson emerges from a balanced action, they do not need anti-bosons, or else fermions would need an infinite number of bosons to exist. To further understand this fundamental phenomenon, refer to the first law of existence stated in the introduction.

Multiplication theory equation explanation

In Newtonian mechanics, force is equal to mass and acceleration. According to the multiplication theory, which uses a universal definition, force is equal to a change in energy, matter, or space. This definition was adopted for the multiplication theory to ensure that we have considered all the possible independent effects that force can cause anywhere it applies in the universe. Specifically, the independent effects (×) and (+) are used in the multiplication theory equation (MTE) because the multiplication theory focuses on force in the aspect of creation.

Note that the inclusion of "2*a*" in MTE 1 is not saying that the second law of motion is wrong. Rather, it asks: What if the body in question doesn't have mass? Or what if a force-carrier produces two acceleration (Pair production or non-pair production)?

You can think of the multiplication equation as a universal mathematical expression of the three laws of motion, where F describes the first law, $(m_1 \times m_1) / (m_1 + m_1)$ describes the second law, and $2a$ describes the third law.

Using something familiar, let me explain the multiplication equation one more time as it applies to both classical and quantum mechanics. An explosive goes off in a building and bomb technicians are called to investigate what happened. When they arrive at the scene—after everyone has been secured and the area is rendered safe to enter—the first thing the bomb squad will want to know is the amount of force released by the explosive. This will tell them

if it was high-powered or just a homemade IED. Thus, the first thing they usually do is measure the distance covered by the force of the explosive.
How does a piece of matter the size of a cell phone transform into a huge ball of energy that travels over half a mile? The question here isn't actually the question of distance or acceleration, even though it's equivalent to it. Rather, it is a question of how a small piece of matter changed into a large amount of energy. This question can be compared to the multiplication effect ($m_1 \times m_1$) in the multiplication equation. Under the influence of a force, it is the highest potential effect that could happen to the body in question.

The second effect that could reveal the strength of the force is how things fuse together at a crime scene. For instance, if you find a pen in its original shape embedded in a dumbbell at the scene where an explosive went off, or you see a refrigerator that now looks like a flat-screen TV after slamming against the wall, you don't need Galileo to tell you that you are dealing with a high-powered explosive. This knowledge can be compared to the addition effect ($m_1 + m_1$) in the equation above. Under the influence of a force, it is the strongest potential effect that could happen to the body in question.

Why didn't I examine the broken glass or matter (m_1 / m_1 *or* $m_1 - m_1$) in the building to determine the amount of force yielded by the explosive? To answer this question, let's shift our crime scene to the quantum world.

You arrive at a scene where a bomb has just been detonated and it's your job to determine how much force it released. But guess what? Everything in the building, including the building itself, is a fermion. If you are a physicist, you know it doesn't matter if the bomb that went off was a supernova—fermions cannot be divided further. Thus, instead of looking for broken fermions—($\psi_1 - \psi_1$) or ($\psi_1 \div \psi_1$)—you would be looking for other clues to determine the amount of force yielded by the bomb. For example, how fermions change in appearance ($F = \Delta$), how two or more fermions fuse together to form another particle ($\psi_1 + \psi_1 = \pi^+$), or how two fermions collide and produce a high-energy boson, starting a particle chain reaction ($\psi_1 \times \psi_1 = \gamma, \gamma$), ($\gamma = 2\psi$).

If you find an electron-positron or up quark-antiup quark in this crime scene, you could predict that two opposite charged fermions had collided (multiplied). If you see a bunch of photons in this crime scene, you could predict that two opposite electric-charged fermions had collided (multiplied). If you find mesons from the first generation of matter, you could predict that mesons from the second generation had multiplied, leaving behind mesons from the first generation. But if you detect a top quark in the crime scene, you can conclude without a doubt that two multiplicative powers, such as proton, had instigated the act of multiplication.

The addition sign in $m_1 + m_1$ represents the presence of a strong force, while the multiplication sign in $m_1 \times m_1$ represents the presence of a high-energy interaction. The probability that one of these effects occurs when force is applied can be derived by dividing the strongest independent effect by the highest independent effect $(m_1 \times m_1) / (m_1 + m_1)$.

For a massless particle, the entire energy (E_1), equivalent to its kinetic energy, can be used to represent it. The only difference is in quantum mechanics, the reaction of force mediators leaves behind two waves, which is unlike classical mechanics, where the reaction of force often causes acceleration. These waves are denoted by 2 in this equation: $K_F = [(E_1 \times E_1) / (E_1 + E_1)]\ 2$.

As far as we know, massless particles are usually force carriers, and as such, they have a constant speed regardless of their energy level. Speed is related to force and kinetic energy. Hence, a body traveling at a constant speed will have a constant kinetic energy: $K_F = [(E_1 \times E_1) / (E_1 + E_1)]\ 2$.

Unstable existence

After examining every known element on earth down to the very atoms that make up all observable matter in the universe, it is without argument that all stable matter consists of either a multiplicative reaction (½), a multiplicative power (≥3), or the combination of both (3 + ½). One of the reasons mesons, described as $\underline{2}$, are unstable, can be understood through simple mathematics, as seen in the Hadfer table. The number $\underline{2}$ is a mathematical sign that describes the combination of equal and opposite reactions of force.

Let me start this explanation by stating that the reactions (-½, ½) of a boson (1) can never come together to form a stable particle. This is because they are two incompatible reactions. Compare this to how each broken side of a magnet repels after they separate.

A gauge boson can be compared to a magnet. Once a boson (1) becomes a separate reaction ($\underline{2}$), unless you are planning to blow something up, don't bring two equal but *opposite* reactions together. (This analogy should not be construed to mean that a boson *splits* into two reactions. The reactions of a boson are more like the energy of the boson released as two equal but opposite energies.)

Opposite reactions of force are naturally incompatible; that's why particles and antiparticles annihilate upon contact, according to the multiplication theory. Accordingly, in both classical and quantum mechanics, when the result derived from the calculation of force is equivalent or similar to an equal and opposite reaction *alone*, the result will be unstable. Starting from zero, if you add or multiply a number by itself, the only numeric and non-numeric values that produce an equal result with opposite, independent arithmetic operations are 0 and 2.

Let's take a simple problem as a mathematical demonstration to explain how the number 2 describes instability. This example does not say that anything with a mass of 2 kg is unstable. Rather, it uses the number 2 to *describe* a process where the opposite reaction of force interacts. As you've learned previously, compare this math to a dream.

During a golf game, a golf ball with a mass of 2 kg rolls toward a hole at a speed of 5 m/s. Using the Newtonian equation, the force applied to the golf ball would be calculated as:

$F = ma$

$F = 2\text{ kg} \times 5\text{ m/s}$

$F = 10\text{ N}$

Notice that 5 m/s is just one part of the reaction (equal reaction). The opposite reaction is not included in $F = ma$.

Now suppose the equation is expressed properly to include the effect on mass as seen in the kinetic energy equation ($\frac{1}{2}mv^2$) and the third law of motion. The force would be calculated as:

$F = \frac{1}{2}m\ (2a)$
$F = 1\ kg \times 10\ m/s$
$F = 10\ N$
Or, according to the multiplication theory equation:
$F = [(m_1 \times m_1) / (m_1 + m_1)]\ 2a$
$F = [(2\ kg \times 2\ kg) / (2\ kg + 2\ kg)]\ 2 \times 5\ m/s$
$F = 1\ kg \times 10\ m/s$
$F = 10\ N$

In classical mechanics, when force is calculated, any action that produces an equal and opposite reaction alone, leaving behind no traces of the mass, is explosive. Apparently, the golf ball above is not described as a bomb.

In quantum mechanics, when a fermion and antifermion bind together, such as in mesons, even though the particles are not equal, the implication of the interaction between opposite reactions gives rise to particle instability.

This incompatibility is more apparent when using the kinetic energy equation. Let's calculate the kinetic energy (E_K) of the golf ball in the problem above.

$E_K = \frac{1}{2}mv^2$

$E_K = ½(2 kg) (5 m/s ×5 m/s)$
$E_K = 1 kg × 25 m^2/s^2$
$E_K = 25 j$

Notice from the results of this equation and the previous one that it is as if the golf ball doesn't have any mass at all.

Minimum Effects of Force

The evolution of the atom, starting with the fermions to the molecules of every element known in the universe, rather than resulting from a fine-tuned design by a supernatural deity, went through all possibilities.

When all possibilities are inserted into a question, the answer depends on the limit of the possibilities. If the possibility is one, the answer will remain the same; but when the possibilities are more than one, the answer is equal to all possibilities.

For instance, anyone who thinks a Feynman diagram is difficult to understand hasn't learned the basics of elementary particles. And anyone who thinks a Feynman diagram is *not* complicated hasn't considered all possibilities during particle interactions.

I'm reminded of a famous quote by Albert Einstein: "God doesn't play dice with the universe." I agree. God doesn't play dice. That's because God *is* the dice, and we are the gamblers. The dice are not confused; we are. Even though the dice can only have one possible outcome at a time, this outcome was already determined the moment they were tossed. As long as speed and direction remain constant, you will get the same result every time you toss them.

The reasoning that led to Einstein making this statement emerged from his belief that gambling contradicts the idea of an all-knowing God. However, as I have discovered, rather than gambling, the universe operates on the principle of freedom.

Imagine if God made every creature perfect and submissive to love and obey him without the ability to make independent decisions. Like a captor who has forced a woman into marriage, God would never experience real love. Thus, a creator who wants to experience unconditional affection would allow his creatures to have the freedom of choice.

Even if such a Creator had the power to create a one-sided die, he would add more sides to it. You may call this gambling, but I call it freedom of choice. If all God gave us was love, we would have no choice but to love him. Instead, he gave us the choice to love or hate.

You may not get as excited when you are about to watch a movie you've seen before. Occasionally, there's joy in not knowing what is about to happen. In this sense, an all-knowing God might intentionally put himself in an avoidable mystery by creating multiple-sided dice. A real choice is not one that is forced upon a creature. A real choice is one that a creature is allowed to make without obligations.

When I studied the role natural freedom played in creation, I noticed that without the influence of a divine designer, fermions would have gone through every possible effect before the atom was formed, and the evolution of different particles would have happened in the same way.

The fermions that make up hadrons came into existence individually. What happens to them afterward depends on the effects of force. How fermions evolved into various particles through different effects can be understood with this next scenario.

Assume your name is Universe and you are in the business of printing money. Money here represents fermions. For safekeeping, you need to deposit your money in the bank, so you open a business account with every bank in town. In this scenario, banks represent force, and the town represents space. Here are the names of all the banks in town: BOS, Bank of Subtraction; BOA, Bank of Addition; BOD, Bank of Division; and BOM, Bank of Multiplication. The amount placed in each account will be deposited according to the name of the bank.

The first year, you deposit $0. At the end of the year, all your accounts will be empty.

The following year, you deposit $1 into each account. The dollar amount here describes a single fermion. At the end of the year, the BOS account will be empty, the BOD account will have a balance of $1, the BOA account will have a balance of $365, and the BOM account will have a balance of $1.

The same principle applies for the next year when you deposit $2 each day, and the year after that, when $3 is deposited each day.

If you have ever wondered why all the matter we see in the universe is full of hadrons with three quarks when exotic hadrons are possible, it may seem as if the universe used a divine power during creation. The universe doesn't need a supernatural designer to accomplish mathematics. The intelligence nature of the universe *is* mathematics.

The same amount was deposited into each bank account, but the differences you notice when you check the balance are based on the effects of each force. Naturally, the universe could produce hadrons with more than three fermions. However, you will see more hadrons containing the number of fermions where stable hadrons begin. This is the point where the binding force needed for a stable hadron is at its minimum. It costs the universe less energy to make a hadron with three quarks than to make one with four, five, or more quarks.

Compare this to the principle of least action, where a particle under the influence of force will take the shortest route that minimizes the action. This is an idea that baffles most people at first glance. How does a particle know which direction to travel? The truth is, like those bank accounts, the wave of the particle actually takes all possible routes, but you see it emerge from the path where the action is minimized.

Chapter 6

Origin of Matter from Nothingness

I believe that the big bang theory, which has been widely accepted by the majority, is not properly structured. Albert Einstein, one of the physicists who initially opposed the theory, once said, "I want to know how God created this world. I'm not interested in this or that phenomenon, in the spectrum of this or that element. I want to know his thoughts. The rest are details."

If, like myself, you don't understand much physics, but like Einstein, you wish to know how God created the world, let me answer your question about existence with the multiplication theory. How did everything you see around you come into existence? The multiplication theory says everything started with the first law of existence. The beginning that had no beginning in this law *could* mean that the nature of "nothingness" is a hybrid of space fabric, and quantum fields that can neither be created nor destroyed. If proof can be presented, this beginning could also mean a supernatural God or the Lemaître primeval atom. That's why I emphasized the word *could* in the definition above.

The compositions of this hybrid that can neither be created nor destroyed are divided into three fundamental groups: fermion fields (first effects), a space field (first occupation), and boson fields (first causes). Fermion fields have one assignment: to provide a tangible body for matter. A space field has two assignments: to provide occupation and direction for matter. Boson fields have three assignments: to create wave particles in fermion fields, build bigger particles with fermions, and change or break down particles made of fermions.

In this theory, as mentioned earlier, the force-carrying particle that binds fermions together is represented as addition. Accordingly, if you add three of these fermions (quarks) together, you'll get a particle called a proton. Add three more quarks together and you'll get another particle, a neutron. Add neutrons, protons, and electrons (single leptons) together, and you'll get an atom. Add atoms together, and you'll get a molecule. Add molecules together, and you'll get gas, dust, water, me, you, your dog, your grandma, stars, galaxies, asteroids, and all observable matter in the universe.

If you are wondering why those fermions that were added have to be three and not two, this theory says that a stable existence is mathematically represented by a number that allows the result of addition and multiplication to be different. In conclusion, it says that if you bang the nature of nothingness with a multiplicative force, something tangible will come out of it. And the greater the multiplicative power you use to energize a multiplicative force, the greater the tangible particle you will produce in a fermion field.

One of the greatest and most puzzling questions ever raised is: What are energy and space made of? I must admit, when I was young, I used to think that God created everything (even though I didn't understand what God meant). As I got older, this thought gradually faded away as I increasingly realized it would be impossible for God to have created "everything" (not as it is defined in this book) if he is also part of everything. Especially, I realized, God couldn't have created his own occupation.

For a long time, I pondered over what the real explanation to existence would look like until I thought to myself, What if the invisible nature of God, which fills heaven and Earth, is what we refer to as energy? What if God is an occupation? Only then did everything start to make sense.

After many years of probing the mystery behind existence, every logical reasoning I extracted from what I know, think, and believe pointed toward three resolutions, which I called the laws of existence. With these laws, I have been able to come to the conclusion that energy and space are the fundamental nature of nothingness, without which, existence would not be possible. This fundamental nature also means that space and energy can neither be created nor destroyed. As a result, matter owes its origin to the interaction between space and energy.

In a theoretical sense, this interaction can be broken down into three words: momentum, force, and change. Quantum fields are more observable than space fabric, which means they have more density. The density of quantum fields gave rise to momentum, which is a natural propensity to undergo motion without external force and as such, gave rise to the path leading to observable time.

The tiny momentum exchanges between fermion fields, however, aren't enough to start observable time. Force is needed. This is where boson fields step in. As momentum interactions between fermion fields may not be able to instigate observable fermion field oscillation, they are enough to start observable oscillation in a boson field.

Boson wave particles (observable oscillation) are force-carriers. But at the end of their life, think of their exit as the only hammer capable of causing "observable" oscillation in fermion fields. Since energy cannot be destroyed, this "exit" or "end of their life" implies a reaction where the gauge boson changes from one form to another by way of multiplication. When this hammer or force-carrier reacts in fermion fields, it does it according to the third law of motion. The reactions left behind by bosons are what we see around us as elementary particles of matter.

Let me write a mathematical description of everything I've said so far, beginning with a quote from a well-known physicist.

"*If you can't explain it simply, you don't understand it well enough.*" –Albert Einstein

How did the universe begin? = FLE
FLE means "first law of existence"
The non-created beginning in FLE = intangible fields
Space field = $\tilde{O}$
Quantum field momentums = $\underline{0}$ and $\bar{o}$
Boson = $\underline{0}$
Fermion = $\bar{o}$
$\bar{o} + \bar{o} = \Delta\underline{0}$ (exchange of field momentum)
$\Delta\underline{0} \times \bar{o} = \Delta\bar{o}c^2$
$\Delta\bar{o}c^2 = 1c^2$
$1c^2$ = two matter particles (mc^2)
(mc^2) = $2\psi c^2$ (pair production)
2ψ = fermion and anti-fermion
ψ = fermions such as electrons or quarks
ψ^+ = antiparticles, such as antiquark or positron
(+) = gluons

$\psi + \psi^{+}$ = mesons
$\psi + \psi + \psi$ = hadrons, such as protons and neutrons
One or two hadrons + one electron = atom
Energy = atom and atom = energy ($E = mc^2$)
Universe = energy, matter, and space
Energy, matter, and space = everything
$\Delta E/h \; \alpha f = 2\psi c^2$
$F = ma$
$F = [(m_1 \times m_1) / (m_1 + m_1)] \, 2a$
$K_F = [(E_1 \times E_1) / (E_1 + E_1)] \, 2$

At the beginning of this book, I argued the possibility of the primeval atom existing without space. I based this argument on reality and the definition of matter. Using the process of elimination, I'm going to start by pointing out two problems in the presentation of the big bang theory and move forward from there.

Problem 1. Based on how the big bang theory describes the beginning of space-time, you should ask, "Did the big bang expansion or explosion release ready-made particles of matter that formed the first protons and neutrons, including the ones we see today, *or* did the explosion or expansion give rise to high oscillations in fermion fields where elementary particles of matter are created?"

I can't imagine a physicist claiming that quantum fields were created from the explosion or expansion of a primeval atom when we *know* the subatomic particles that make up atoms originated from quantum fields.

Problem 2. According to the big bang theory, all the particles of matter we see today were once condensed in a tiny point about the size of an atom or smaller. In case you are not familiar with the origin of particles, when particles of matter, such as fermions, are created, an opposite is created simultaneously.

It is a well-known fact that a particle and its antiparticle annihilate each other upon contact. This is because equal but opposite reactions of force are naturally incompatible, as explained in the multiplication theory. For this reason, particles and antiparticles couldn't have existed together in a single point. If space didn't exist around the primeval atom, the antiparticles of the atom, when they came into existence, would have been forced to collide with the particles, as they would have had nowhere else to go.

Upon reading this claim, it would be reasonable for you to think this annihilation of particles and antiparticles of the primeval atom is why the big bang occurred. I agree. I wish this were how the tiny, hot ball of matter was described. Think about the logical sense the big bang theory would make if the explosion or expansion were described as the annihilation of the first atom and antiatom in the universe.

However, if this were the case, the question about the primeval atom would then be: Did the existence of subatomic particles precede the existence of the quantum fields where they are created? And *where* did the annihilation happen?

If there's one thing Richard Feynman established during his lifetime, it was that fermion and boson fields work hand in hand to bring particles of matter into existence. Accordingly, the story of the universe is not so simple, though it is quite predictable. It is a story that was once told by many philosophers, some of whom weren't lucky enough to share their thoughts properly before being condemned by outraged religious believers. One, in particular, was Giordano Bruno, an Italian philosopher and mathematician who was sentenced as a tenacious heretic by Pope Clement VIII and was burned at the stake. This fear of condemnation would have its impact as, after the atrocious punishment imposed on Bruno, no reasonable philosopher would dare to claim the universe had no beginning.

Fast forward to 1931, when Georges Lemaître published an article on his theory of the primeval atom. Given the history of mankind and their beliefs, it's no surprise the universe had a beginning in Lemaitre's theory. This idea seems to sit well with many, including religious believers. This theory, where the explosion of a single atom and the expansion of space happened in the middle of nowhere, would later be known as the big bang theory.

Let's reason together. If you have ever worked in a profession that requires incident reports, you have probably learned the five questions that are necessary to write them: what, where, who, why, and when. Out of these five questions, *where* the incident happened must first be stated in every report to make sense of the remaining questions.

Without a crime scene, there can't be a crime. Without a nation, there can't be a president. Without occupation, the big bang couldn't have happened as hypothesized. If you're thinking, "Maybe the author of this book doesn't understand that the big bang created space," I disagree with you. A universe that follows reasonable fundamental principles like we see today couldn't have had an unreasonable beginning. I agree; common sense is not always correct. But I refuse to accept the hypothesis claiming a crime created the crime scene. If the big bang is compared to the first crime, it simply means the first crime scene (space) was not created.

Before the first telescope was built, before man even knew that atoms existed, when our religious beliefs embedded in us the thought that *everything* was created by a supernatural being, there was no science-based reason for us to think that *emptiness* or empty space (the very nature of nothingness itself) was created.

When taking into account the first law of existence, it is easy to see that the universe doesn't need a creator or designer—no divine God, singularity, primeval atom, or big bang theory. Nothingness (quantum fields and space) has always existed and will forever be. The question of how the universe began is an unnecessary burden that religious thinking brought upon humanity. And the main reason why the big bang hypothesis became more acceptable to the public was because the pope at the time believed it matched the idea in the first book of Genesis.

The notion that the universe (space and time) came out of absolutely nothing (absence of space) is not a reasonable thought, and under no circumstances can it be manipulated into becoming one. It doesn't matter how many times a genius advertises the idea—the mind of a human cannot *truthfully* make sense of the incompatibility in the statement, not even those who believe the idea to be true. At one point in the future, I believe humans will come to realize and accept the true nature of nothingness as space and quantum fields. There's an extent to which an unreasonable idea can push the brain. Like a gunpowder explosion, it could be strong in the beginning, but eventually, it will let go of the bullet.

Cosmic inflation

On the idea of cosmic inflation, my respect goes out to everyone who has endeavored to explain the mystery behind our existence, whether they were wrong or right. The selflessness and curiosity of these men and women have made the world a better place. Humankind will forever be grateful to these scientists.

But the fact that anyone would claim that Newton's concept of gravity, which has no evidence of attraction, could act as a repulsive force responsible for the hypothesized inflation shows how we are struggling with our logic. The big bang theory claims that space didn't exist before the singularity expanded, so I doubt that the repulsive gravity used to describe the inflation is referring to the curvature of space.

The more we champion thoughts that have not yet been proven, like wormholes, gravitons, weakly interacting massive particles (WIMPs), time travel, parallel universes, etc., the more I realize that modern science has drifted so deep into the realm of science fiction that we can now create explanations to back up anything we want to fit. We can't yet explain how Newtonian gravity attracts, and we are already introducing a repulsive force of gravity. We have yet to prove string theory, and we are already introducing the concept of a super-string. We are still struggling to understand the origin of our own universe, and some physicists are already introducing the idea of a multiverse.

Origin of matter

I will now explain how quantum fields could have gone from unobservable to observable through a gradual process of addition and multiplication. In this hypothesis, I will describe the origin of observable time in the universe, not space. The sequence of events I will mention is reasonable yet hypothetical. However, anyone who has studied quantum mechanics can theorize that the origin of matter evolved gradually, just like the origin of species in biology.

In my opinion, the process of creation that happened in space was simple, but it requires wisdom and critical thinking to decode. It undeniably began with density and momentum (gravity).

Momentum and density, which I often refer to as *freedom*, is how it all began. Imagine space without observable time—a dark, infinite space where the oscillation of quantum fields is nearly immeasurable. This Planck time, as most physicists would describe "nearly immeasurable," or "quantum foam," as physicist John Wheeler (1911–2008) referred to quantum fluctuation of space-time on a tiny scale, wouldn't last forever, as it would gradually give rise to observable time after a long period of fermion fields exchanging tiny momentums. One Planck time may be too small to measure until it is added to another Planck time.

Gradually, at every point in space, two tiny oscillating quark fields gave rise to high oscillations in a gluon field. As gluon particles emerged and reacted, quark particles began to surface little by little. Gluons bound them together faster than they could be observed individually. Steadily, the numbers multiplied, and over billions of years, large numbers of protons and neutrons appeared in the cosmic atmosphere.

The initial cosmic atmosphere was dominated by quarks and particles composed of quarks (hadrons) for a long period. However, hadrons reduced at the rate they were produced because the atmosphere was in a constant battle between baryons and anti-baryons. On the other hand, electrons and neutrinos kept showing up in proportional quantities as neutrons and mesons decayed, and protons and antiprotons collided, leaving behind a trail of multiplication.

At one point, electrons and neutrinos in the cosmic atmosphere outnumbered protons and neutrons, since they are produced from decay; thus, they have little or no antiparticles to annihilate them. The same applies to quarks produced from W bosons when protons and antiprotons collide.

I don't agree with the Q-balls hypothesis, which is an attempt by some theoretical physicists to explain what happened to all the antiparticles that were released by the big bang. Instead of looking for a more appropriate answer to what we don't know, we do damage to physics when we bend what we *know* to be true to give way to what we *assume* is true. We *know* particles and antiparticles are produced in the same proportion. Thus, any theory that can't explain the lack of this proportionality needs to be double-checked.

During this period of constant particle reactions, annihilations, and decay, apart from quantum fields, which are producing fermions, and gluons (addition) binding quarks together, particles were also produced through multiplication when existing particles collided. When electrons—which were everywhere at the time—interact in what is known as *electron scattering,* photons are produced. Photons are stable particles and have no anti-photon; thus, their production increases in quantity as well.

Over a long period of time, the cosmic atmosphere was eventually saturated with cosmic radiation from the constant annihilation of particles and the abundance of electromagnetic waves.

The production of atoms happened similarly to that of hadrons, slowly and gradually. The minimum effect of force—or principle of least action—started to become obvious, as one of the first atom was only composed of one proton and one electron. This is the minimum effect of force needed to produce an atom. This atom is today known as the *hydrogen-1* (Protium), and it dominates the percentage of observable matter in the universe.

It's okay to agree or disagree with the hypothesis I have put forward. None of us can give a perfect theory of the early universe, as none of us were there when it happened. The best we can do is hypothesize based on what we know to be true. Anyone who has studied physics can theorize about what happened after hydrogen, helium, and lithium atoms formed. Thus, we can fast-forward to the first star or stars in the universe.

The first supernova in the universe would play a huge part in explaining most of the observations in our observable universe today. Perhaps the very first massive supernova or black hole collision, which probably occurred around 13.8 billion years ago in our observable universe, is where the big bang theory should begin, if it is willing to get rid of the creation of empty space. If the actual age of a star known as the Methuselah (as of 2022, it is the oldest star in the universe) is unbiasedly studied, I strongly believe there must be a star out there in the universe older than the date of the big bang.

Chapter 7

Principles and Behaviors of Expansion

It's possible that a similar situation to the one I'm about to describe has happened to you. I'd just left my sister's house after a short visit. About three or four miles down the road, I noticed that my cell phone was not where I usually placed it in the center console of my car, so I headed back to her house thinking I must have left it there.

I rang the doorbell and she opened the door. She wasn't expecting me to come back, so she asked what had happened. I replied, "I think I left my phone here. I couldn't find it in the car." She looked at my hand, and at the same time, I did too. Guess what I was holding. My phone. It had been in my hand the whole time.

That cell phone wasn't invisible. The reason I couldn't see it was simply that I didn't *suspect* it was what I was holding. I want you to keep this story in mind as you continue reading.

I define *expansion* as a type of change that causes an increase in size or distance on the shape, structure, or influence of something.

The word *expansion* can be used in many ways, most of which may not relate to a physical thing. For instance, I can say, "My business is expanding." Even though my business is an online software company that has no physical structure in real life, "expanding" perfectly describes the increase in the influence of my business on the internet.

However, when expansion is happening to an observable form of existence, such as energy or matter, it must obey three principles.

First principle: Force is required for expansion to begin.

Second principle: Change must be observed to account for expansion.

Third principle: Provided a force remains constant, change must be proportional to the force.

When expansion is happening to an invisible form of existence, such as space or on a surface where motion is not instantly noticeable, something visible or observable within the entity is needed to prove the expansion. For this visible or observable thing or things to prove that expansion is happening, these are the changes or behaviors they must exhibit:

Behavior 1: The measurement, shape, or size of the observable thing or point in any dimension of space must increase.

Behavior 2: If an equal vector force is behind the expansion, one observable thing in any dimension of space will be stationary, while another observable thing in the same dimension will move away in one direction, and at a speed directly proportional to its density.

Behavior 3: If an equal and opposite vector force is behind the expansion, two observable things in the same dimension of space will move away in opposite directions, and at a speed directly proportional to the product of their densities.

Behavior 4: When more than two observable things in one dimension of space describe expansion, the motions and speeds of these things will not be uniform, and compression of space will occur between them unless the force behind the expansion is concurrently acting on all the surfaces of the dimension or outside the space between the things.

Behavior 5: When there are more than five observable things, at least one in both directions of each dimension of space that describes expansion, the motions and speeds of these things in all dimensions will not be uniform, and compression of space will occur between them unless the force behind the expansion is concurrently acting on all the surfaces of the dimensions or outside the space between all the things.

Expansion demonstration

Similar to how we follow two vectors (e.g., vertical and horizontal axes) before drawing a Feynman diagram in physics, you can demonstrate expansion by drawing three double-headed coordinate arrows to represent three-dimensional space. The double heads represent the opposite vectors of each dimension of space.

Space isn't something you can see or touch, so when space expands, you will need something observable to detect the expansion. This is where time comes in. We need time to be present in the dimensions to make sense of the symmetrical nature of space. This is why time is regarded as the fourth dimension in general relativity.

Figure 1

In Figure 1, "BB" stands for "big bang."

By using something observable, such as dots, to represent galaxies, try to demonstrate the expansion of space (the arrows). Keep in mind that according to the big bang theory, galaxies weren't there at the moment of the big bang; thus, in the demonstration of space-time expansion, the observable time (dots) cannot begin at the same time as the expansion.

Firstly, if you are willing to be unbiased, in a space that is expanding at almost the speed of light, you should wonder how galaxies are able to form. But let's agree with popular belief that a weak force of gravity was able to overcome this intense expansion. If you perform this demonstration with an open mind, you will notice a pattern that corresponds with one of the principles and behaviors mentioned above.

It's easy to demonstrate expansion with well-arranged galaxies on a two-dimensional space. If this two-dimensional demonstration of space-time expansion is accepted without rigorous questions, one might be convinced that the universe is expanding uniformly, as described.

For the purpose of a three-dimensional demonstration of space-time, let's assume the universe has a center. Except at the center of the universe (BB), you can start the demonstration by putting the Milky Way anywhere on any arrows in Figure 1 and apply this statement: "It doesn't matter where we point our telescope—galaxies are moving away from us in all directions." Put as many dots as you please on any of the arrows to represent the receding galaxies in this statement.

If you compress the arrows in Figure 1 to a single point (BB), the only way to stretch them out uniformly in all directions is by pulling on the edge of the circumference or pushing out the surface concurrently at every point. The force applying the pulling or pushing could either be inside or outside the circumference. Either way, the force needs something to pull or push against for expansion to begin, $F = ma$.

Based on this fact, I concluded that a force that started from a single point and stretched out the three dimensions of space we see today would have required space to have an edge in all dimensions.

In 1929, Edwin Hubble claimed that the majority of the galaxies he observed were receding from our own in all directions. I believe he did not make up what he saw. But keep in mind that the Milky Way is not the center of the universe. Thus, if you agree that our galaxy is not lucky enough to be located at the center of the universe (that is, if the universe even has a center), the conclusion that all matter in the entire universe must have once occupied the same point is a preconception. The more closely I look at this conclusion, the more I realize our logic still bears traces of our old misconception that we are at the center of the universe.

Assuming I agree with the expansion of empty space, I'd still argue that just as the stretching of empty space leaves behind more empty space, a compressed empty space would also leave behind more empty space. This is because space is intangible, and therefore, the absence of empty space can only be described, defined, or demonstrated by the presence of something tangible or observable.

Space and empty space, as I defined them at the beginning of this book, can be compared to light and darkness. The only way to remove one is by introducing the other. Therefore, it is impossible to demonstrate the total absence of space and empty space.

I know we can't depend on common sense alone to judge everything we observe, yet our science was founded upon the guidance we receive from our common senses. When it comes to the big bang theory, we have been told to ignore all the questions that our senses and mind tell us to ask, such as, what is the universe expanding into? Where did the big bang happen? How did the primeval atom exist without space? And what created the atom?

If you have asked any of these questions, I implore you to agree with your innermost instincts telling you they must have a reasonable answer. If science cannot offer reasonable theoretical framework to answer these questions, then it is not yet in a position to judge anyone for believing that a supernatural God exists.

Rather than expanding, I expect *if* a light wave or any quantum wave is in fact stretching, the ripple in space should be used to explain the stretching.

Albert Einstein became famous in a matter of days because his prediction about curved space was confirmed with the bending of light observed during a solar eclipse. The idea he theorized prior to this confirmation says that if space is bending around the earth, light, which is a property carried by the fabric of space, should bend around the earth as well. If the motion of light according to this idea is controlled by the motion of space, in theory, it means if space oscillates like a wave, light traveling through it would oscillate like a wave as well.

Accordingly, we can say that when the wavy motion, or ripples of space, is stretching out like ripples would do after a while, a light wave traveling through it would stretch out as well.

If you disagree because you believe quantum waves are independent of space's motion, wouldn't that contradict your belief that the expansion of space stretches the wavelength of light?

Based on a direct, demonstrable experiment, we are more confident about ripples of space than the expansion of space. Therefore, *if* light waves are in fact stretching due to space, I expect the Doppler effect, or redshift, to be as a result of ripples of space (proven) rather than the expansion of space, which has no direct experimental proof.

According to how the Doppler effect was explained by Christian Andreas Doppler (1803–1853), the frequency of sound or light waves depend on the speed and direction in which the source is moving relative to the observer. This is why we could demonstrate the Doppler effect—stretching or compression of a light or sound wave—when an ambulance passes a person standing on the side of the road. Never before had anyone concluded that the road is expanding until the big bang theory needed to borrow the idea.

Let's agree momentarily that "all" galaxies are stationary. Even if you are not an astronomer, at least from general relativity, you must have learned that all matter in space is undergoing a free-fall motion. This alone should raise suspicion when someone claims that galaxies are stationary and their motion is due to space increasing between them.

Think about it: How do "supposedly stationary galaxies" beyond the gravity of each other head toward one another until they collide? When half of what our telescopes see is used as evidence for the big bang theory, the remaining half is either not so famous or unintentionally omitted when discussing the big bang evidence. One example is the Andromeda galaxy, which is currently heading toward the Milky Way galaxy.

One cannot say astronomers intentionally refuse to make merging galaxies as famous as receding galaxies. When someone believes something to be true, anything that contradicts this belief could often be seen as less important. Especially when astronomers believe gravity could explain why galaxies collide. But could it? Could a weak force whose strength is less than the width of a proton be responsible for merging galaxies?

In a universe where all galaxies are said to be stationary but are moving away from one another as a result of expanding space between them, shouldn't galaxy merging indicate a problem with the theory of expansion?

One example of such a merger is NGC 4676, the Mice galaxies, identified by the Hubble Space Telescope. Another two galaxies in the process of colliding are NGC 2207 and IC 2163 (discovered by John Herschel in 1835), also recorded by the Hubble Space Telescope.

These galaxies are just a few examples of the ones we have discovered in the tiny fraction of the universe we can observe. Imagine how many more must have collided in the past, and how many more are still waiting to be discovered.

Suppose when Edwin Hubble was observing distant galaxies with his telescope that all the galaxies he happened to see were merging or colliding.

Do you think he would have concluded that the universe is expanding?

If you grant my argument the benefit of the doubt (just a shred of doubt), it's possible that the outward ripples in space-time, which began around 13.8 billion years ago from the first massive black hole collision, could be what we have misunderstood as *dark energy*. We couldn't see this effect simply because we didn't suspect it, like a cell phone in the hand or sunglasses on the head.

Here is another way to understand the argument against dark energy. The mass of a body, such as the sun, is a contact or applied force that causes space to bend around it. Based on a testable protocol, a force that is needed to stretch a body must be double or greater than the force that is needed to bend it. Scientifically speaking, the energy required to stretch a body must be twice or greater than the energy required to bend it. When both forces act on the body simultaneously, the energy required to stretch the body would overcome the energy required to bend it. Pick any object around you and use it to test this claim. According to my test, I concluded that it is impossible for dark energy and the gravity of general relativity to coexist.

To put it simply, *dark energy doesn't exist as described.* Dark energy is like an error in a manuscript. It doesn't matter how many times an author goes over their own manuscript, there will always be some errors they do not notice.

These outward ripples of space, if observed closely, could resemble an expansion; but they do not need mysterious energy to continue. Again, I would argue against the existence of dark energy based on the first law of motion. The argument is simple: Once the first black hole collision set the ripples of space in motion, a body in motion remains in motion, and a body at rest remains at rest unless acted upon by external forces. Therefore, after a black hole collision starts what seems like the expansion of space, it is unnecessary to hypothesize that a mysterious force (dark energy) is responsible for the continuous outward motion of space.

Dark matter

If you look at Saturn through a telescope, you can't help but admire the beauty of its rings. One unanswered question lurks around this beautiful planet: What keeps its rings together? If you are a physicist and said gravity, you could be right, and you could be wrong. It depends on how much you understand the effects of gravity. But whatever answer you provide to explain the rings of Saturn cannot be completely *excluded* from explaining what keeps the fast-spinning spiral arm of a galaxy together.

Before you roast me with the mathematics of general relativity for making this assumption, let me tell you a short story.

An archeologist from the West embarks on a journey of a lifetime. His plan is to go to remote areas of the earth in search of human tribes that have never experienced modern civilization. At first, the search is boring. Eventually, he finds what he is looking for.

In an isolated part of the Amazon rainforest, one that many archaeologists would describe as uninhabitable for humans, he stumbles upon a tribe as old as mankind. Fortunately for him, they are not cannibals. Although there is a language barrier, they're friendly and welcoming.

The hospitality he receives is so good that he decides to return the following year. This time, he manages to convince the leader of the tribe to follow him back to the West. Of course, they have to board a plane, which isn't as easy as it sounds. You can imagine trying to convince a bushman to enter the mouth of what he thinks is a big, wild bird. But the Western guy manages to get him on the plane, and off they go.

During the flight, something happens. The pilot runs into an unexpected storm. To save the plane from lightning strikes, he has to make a quick decision. Before the plane flies too far into the storm, the pilot turns the plane around by performing a 9-G maneuver. The whole ordeal lasts less than five minutes. In the end, the passengers on board are saved.

Here's where this story is heading. During the 9-G maneuver, every passenger on the plane knows exactly why they are being violently pinned against their seat—except for the bushman, who is convinced that invisible hands are keeping everyone from falling out of their seats. He tries to discuss it with the Western guy, but communication between them is a big problem.

This bushman later returns to the wild, and as time goes on, he is able to convince his people that invisible hands exist inside a big bird that the Western guy referred to as an "airplane."

A year later, when the Western guy returns for another visit, to his surprise, the tribe has crafted the shape of an airplane out of a tree, and every so often, people gather around it to worship the invisible hands.

Now, I know you, the reader, are smart. I believe you know the wisdom I intend to extract from this story. So, let me say it this way: I don't believe "dark matter" exists as hypothesized. But if you believe it does—simply because you can't explain why the spiral arm of a quickly spinning galaxy stays in orbit—I want you to think about the bushman and the invisible hands.

I believe that if someone hadn't described the Higgs particle some few years back; if it had been discovered accidentally; it's possible someone out there might have claimed they'd discovered the dark matter particle. And who are we to believe? After all, the God particle is massive, doesn't interact with photons, and gives mass to the majority of fermions. It's a perfect candidate for dark matter.

Chapter 8
Entanglement

The Theories of Entanglement, And How They Work

Entanglement, as it is observed in experiments, meaning an equal or opposite correlation between two things, even when the objects are separated by vast distances [1.01]. Entanglement is a real phenomenon; that is, it is possible for two objects separated by a large distance to have correlating attributes. The objects involved in entanglement are localized, and the physics behind the occurrence can be reasonably explained.

Three possible processes can cause two objects to be entangled: (1) simultaneous change and time, (2) a fundamental variable within an object that only change when influence by a force, (3) information transfer. Entanglement has been observed between quantum particles like electrons and photons, as well as in classical objects such as diamonds. In the formulations that follow, we shall present two logical explanations of how entanglement works in both classical and quantum mechanics. In addition, we will show that information transfer is not responsible for entanglement in quantum mechanics. Ultimately, without complex mathematical dependency, either one or both of these theories can provide a reasonable explanation of entanglement in both classical and quantum mechanics:

1. The theory of classical entanglement.
2. The theory of quantum entanglement.

Entanglement in classical mechanics

In the study of motion, there are two inevitable effects that happen when force is applied to a body: change and time. Wherever change or time is observed, force is present. One could object to the existence of gravitational force, as Albert Einstein did when he introduced general relativity. Yet, Einstein couldn't help but agree that any change occurring in space can only be understood properly with time. Since time is abstract, it is safe to conclude that it cannot be completely comprehended without observable change. Change, as far as we know, cannot occur without the passage of time. Likewise, the passage of time cannot be perceived without change. Thus, the nature of change and time are inseparable [33.3].

Time is the measure of change. We have defined *change* as it relates to a physical body rather than abstract attributes, as any observable difference in the appearance of something, or a motion that affects the size, direction, or position of a body. The change happening to a body can be fast or slow. The rate at which a change occurs is known as *speed.*

A clock is an example of the inseparable nature of change and time. Change in direction is used to create a clock. However, in order to measure time with a clock, we made sure the motion of the clock hands was a constant speed in a constant direction. We did so because we know the passage of time happens in a reasonable order; that is, the past, present and future cannot occur simultaneously. Therefore, because the nature of time requires that it must occur in a reasonable order, an equal change in a constant direction is necessary to measure time. To prove this, we shall reverse the preceding claim, meaning, instead of using equal change to measure time, we will use time to measure equal change. We will use a clock to measure the change of two identical systems. These systems shall be used to show that two physical objects that have conserved momentums, even if separated by a vast distance, can share a simultaneous reality. The time operator that will be used to determine this correlation shall be referred to with the words *simultaneously*, *equal time, at once,* or *at the same time*. These words shall be used occasionally to mean the same thing. Both systems below have been given a set of equal values, such as equal speed and equal directions.

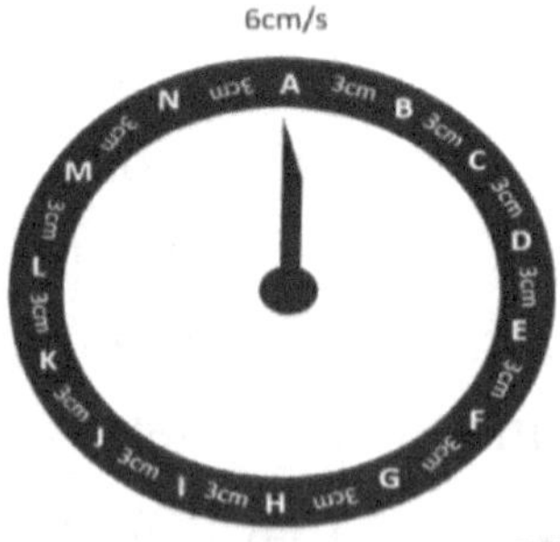

Fig. 417

t = 00:00:00

Fig. 418

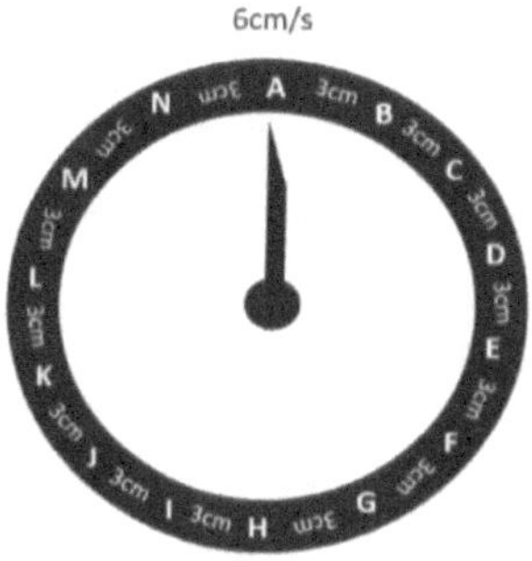

Fig. 419

t = 00:00:00

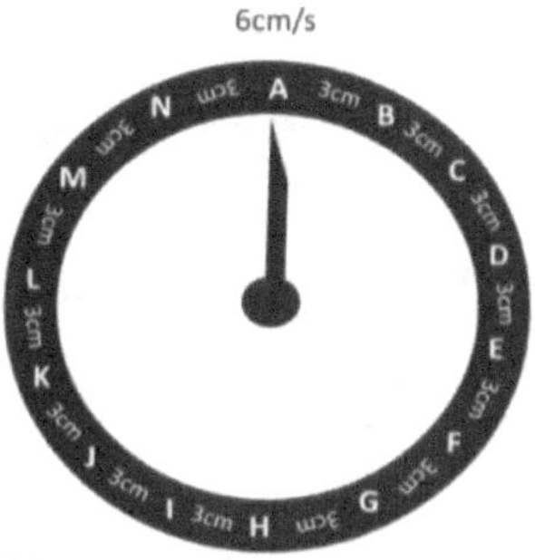

Fig. 420

Figs. 419 and 420 show two identical systems with one propeller each. Both propellers can move clockwise or anti-clockwise. However, in this demonstration, we have agreed that each propeller is identical in terms of mass and they can only change directions with equal speed in a clockwise direction. Both propellers start in a clockwise motion simultaneously from point A with a constant and equal speed of 6cm/s.

Fig. 425

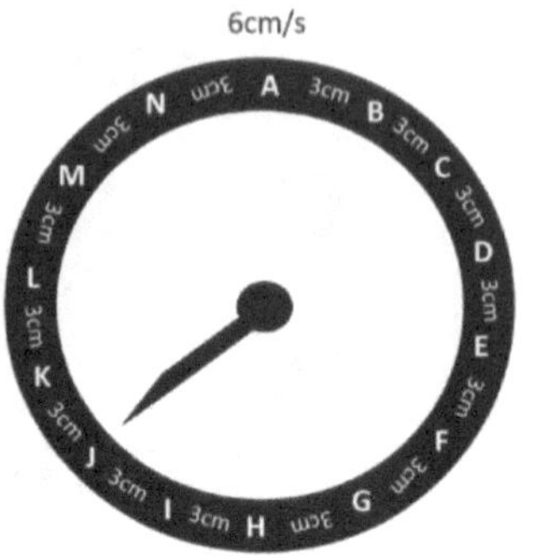

Fig. 426

T = 00:00:01

Fig. 421

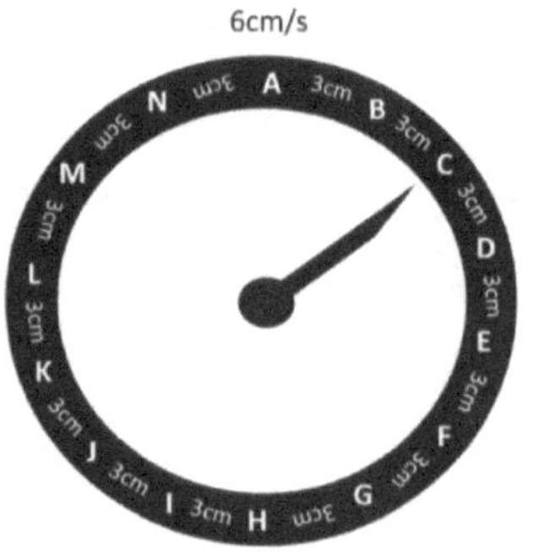

Fig. 422

T = 00:00:01

Figs. 421 and 422 is referred to as an equal oriented correlation. This is because the propellers started moving at equal orientations, as seen in Figs. 419

and 420. As can be seen from the above figures, since the simultaneous change of both propellers are the same in terms of speed and directions, both propellers arrived at point D at equal time (one second).

Figs. 425 and 426 were also stopped at equal time. And because both systems have equal speeds and equal directions, their changes are equal but opposite. Figs. 425 and 426 are considered opposite oriented correlations. This is because the propellers started motion at opposite orientations, as seen in Figs. 417 and 418.

Fig. 430

Fig. 431

T = 00:05:01

Fig. 430 and 431 shows two opposite oriented propellers spinning at high speed. Like the previous diagrams, both propellers have an equal speed. Fig. 430 started in a clockwise motion at point A, while Fig. 431 started in a clockwise motion at point H, simultaneously. In the diagram above, if you pay attention to the digit that counts the seconds, you will notice it appears as if time is superimposed; that is, the past and present are occurring simultaneously. Regardless, I am confident that this is just an illusion caused by digital numbers moving at high speed. You will see the same effect if you start a digital stopwatch.

Also, notice the propeller in both systems. Even though each system contains a single propeller, at high speed, they appear to be in multiple places at the same time. Thus, in both systems above, we cannot determine precisely where the propellers are pointing until they are stopped. A similar effect is seen when a pedestal fan is running at high speed.

However, suppose Fig. 430 is in my possession at point *x*, and Fig. 431 is with my friend Paul, who is located 10,000 miles away at point *y*. If I stopped my propeller at exactly 3:00:00 PM and it points at

G, I'd immediately know that the orientation of Paul's propeller (Fig. 431) would point at *N* *if* it had been stopped at exactly 3:00:00 PM. The knowledge about Fig. 431 is known to me without disturbing the system. The reason why I could accurately predict the orientation of Fig. 431 without measuring it is simply because my propeller and the propeller in Fig. 431 have equal speeds and they begin motion at equal times. Therefore, in two identical systems where all parameters relating to motion are equal, when a mass in each system starts at a given point simultaneously, if at $t = 0$

[1] $m_1 \mathrm{ks}\, \Delta \mathrm{v}\, \mathrm{r} = T_1$ at point D
then
[2] $m_2 \mathrm{ks}\, \Delta \mathrm{v}\, \mathrm{r} = T_1$ at point D.

Mass = m, radius = r, change in velocity = Δv, and, Time = t, T.

ks and Δ in [1] and [2] are indications that mass has a constant speed and velocity has a constant change. If we isolate [1] or [2] without the constants, the multiplication of the remaining values will give us the angular momentum of the systems at a constant time (mvr = T_1). Hence, the conclusion in equation [1] and [2] shows that the angular momentum of both systems is conserved.

If we shift our focus away from the mechanics of motion and observe the correlation between both propellers, you will see that when both propellers start at equal orientations, as seen in Figs. 419 and 420, their changes are equal when we stop them at equal times (see Figs. 421 and 422). Likewise, when both propellers start at opposite orientations, as seen in Figs. 417 and 418, their changes are opposite when we stop them at equal times (see Figs. 425 and 426). Finally, notice that in all the diagrams above, there is no physical connection between the two systems. As a result, because the correlation between the systems is based simply on simultaneous change and time, provided the speed and directions of both propellers remain constant, regardless of the distance between them, both systems will have an equal or opposite correlation when stopped at equal times. According to [1.01], Figs. 430 and 431 are entangled. The way to show that both systems are entangled is by measuring them at equal times. Prior to measurement, any delay in speed or time will destroy entanglement.

In this formulation, we used a specific motion to explain classical entanglement. However, entanglement can occur with different kinds of motion, provided the motions are in accordance with the classical entanglement theory.

Classical entanglement theory

The classical entanglement theory derived from this formulation is as follows: when two equal or opposite oriented objects that have equal speeds begin a simultaneous motion in a constant direction, provided their directions and speeds remain constant, regardless of the distance between the two objects, their change in orientation will always be equal or opposite when they are measured at equal times.

$$(\text{ks } \Delta\text{v} + \text{ks } \Delta\text{v})\ t_1 / t_1 = 2(\text{ks } \Delta\text{v})$$

<u>Opposite correlating entanglement</u>

$$\downarrow\uparrow \quad (\updownarrow\text{ks } \Delta\text{v} + \updownarrow\text{ks } \Delta\text{v})\ t_1 / t_1 = (\uparrow\downarrow)$$

$$\downarrow\uparrow \quad (\updownarrow\text{ks } \Delta\text{v} + \updownarrow\text{ks } \Delta\text{v})\ t_1 / t_1 = (\downarrow\uparrow)$$

$$\uparrow\downarrow \quad (\updownarrow\text{ks } \Delta\text{v} + \updownarrow\text{ks } \Delta\text{v})\ t_1 / t_1 = (\uparrow\downarrow)$$

$$\downarrow\uparrow \quad (\updownarrow ks\ \Delta v + \updownarrow ks\ \Delta v)\ t_1 \,/\, t_1 = (\uparrow\downarrow)$$

<u>Equal correlating entanglement</u>

$$\downarrow\downarrow \quad (\updownarrow ks\ \Delta v + \updownarrow ks\ \Delta v)\ t_1 \,/\, t_1 = (\uparrow\uparrow)$$

$$\downarrow\downarrow \quad (\updownarrow ks\ \Delta v + \updownarrow ks\ \Delta v)\ t_1 \,/\, t_1 = (\downarrow\downarrow)$$

$$\uparrow\uparrow \quad (\updownarrow ks\ \Delta v + \updownarrow ks\ \Delta v)\ t_1 \,/\, t_1 = (\uparrow\uparrow)$$

$$\uparrow\uparrow \quad (\updownarrow ks\ \Delta v + \updownarrow ks\ \Delta v)\ t_1 \,/\, t_1 = (\downarrow\downarrow)$$

Unknown direction is denoted by $\updownarrow$.

Classical entanglement theory test (1.22)

For accuracy in equal time measurements, these tests were performed in two ways, using video motion and computer software. The functionalities in both tests were created based on the classical entanglement theory.

Fig. 33

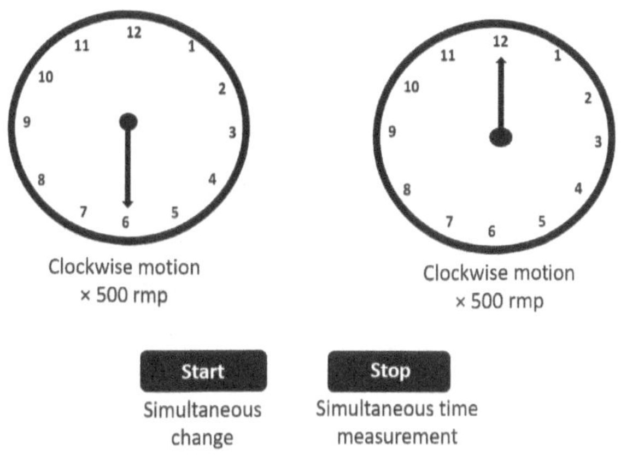

In the video test, the hands of two clocks were synchronized in opposite directions, 12 and 6. Both clock hands were given the same speed (× 500 rpm). With the start button, they began a clockwise motion simultaneously. Using the stop button of the video application to simulate equal time measurements, the clock hands were always opposite when stopped simultaneously.

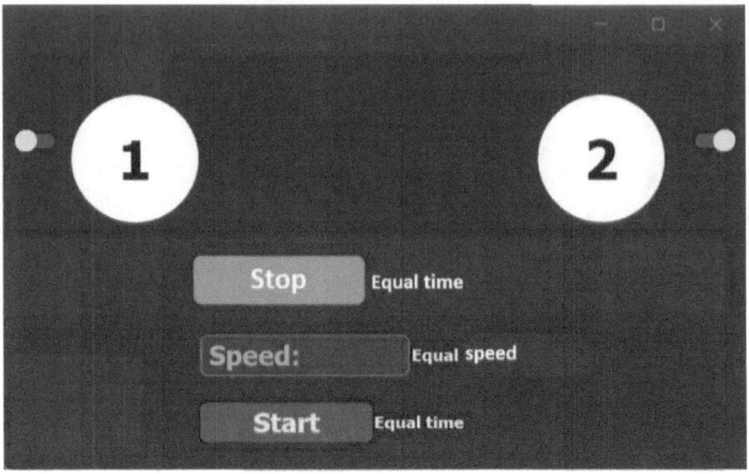

In the computer software test, two numbers, 1 and 2, in each circle, can interchange with equal speeds. The buttons on the sides of the circles are used for changing the numbers. Numbers 1 and 2, or 2 and 1, represent the opposite orientation. The start button was designed to set the numbers in both circles in motion at equal times. The stop button is used to simulate equal time measurement. In other words, it is used to stop the changes of both numbers at equal times. The speed box allows us to enter a high or slow speed. However, both circles use the same speed. When the application begins motion with opposite numbers, each time we press the stop button, the numbers are always opposite. When the application begins motion with equal numbers such as 1 and 1 or 2 and 2, the numbers are always equal each time we press the stop button.

Entanglement in quantum mechanics

The EPR Paper. Authored: A. Einstein, B. Podolsky, and N. Rosen.

Title: Can Quantum Mechanics' Description of Physical Reality Be Considered Complete?

Volume 47. Institute for Advance Study, Princeton, New Jersey. Page 777-780

The EPR paper was authored by Albert Einstein and two co-authors, Boris Podolsky and Nathan Rosen. The objective of the publication wasn't to disprove quantum mechanics, but rather to create a compelling argument showing that quantum mechanics is an incomplete theory. Part of the assertion that led to this argument is a mathematical description in quantum mechanics, which states that after two identical particles interact for a period of time, their wave functions can no longer be considered as separate systems, even if the particles are separated by a distance. For this reason, it is often said that entangled particles shared a *single wave function*. EPR argued that this assertion contradicts the uncertainty principle, where quantum mechanics maintained that the position and momentum of a quantum object cannot be known at once.

The center of the EPR's argument was a thought experiment known today as entangled state or quantum entanglement. The experiment was formulated in this way precisely. Suppose we have a zero spin particle that decayed into a particle *A* and a particle *B*. After both particles separate, they move freely in opposite directions. To conserve momentum, the spin state of both particles is known to be opposite prior to separation. Each of the particles are now separated by a vast distance. If the quantum state, such as the spin orientation of one of the particles, say particle *A*, is measured (determined) instantly, we would be able to predict the spin orientation of particle *B* without disturbing it.

While the entire EPR arguments may or may not be wrong, they've unintentionally uncovered one irrefutable reality of quantum theory: quantum entanglement. This phenomenon has been observed to be true many times in various experiments. Using different knowledge from classical and quantum physics, we will explain one possible method by which two particles can be entangled. In this formulation, we will clarify the misconceptions in some assertions of quantum mechanics and the EPR suggestions. This clarification, which shall be supported by illustrations that are in agreement with empirical observations, will show that two particles, having conserved spin angular momentums, can have simultaneous or delayed correlation. Furthermore, we will identify and explain the so-called *single wave function*, which is supposedly responsible for quantum entanglement. In the formulation of the quantum entanglement theory, we shall touch upon both the accepted concepts and common misconceptions in quantum mechanics. For the purpose of this explanation, an electron shall be used for demonstration.

Superposition of states

Paul Dirac (1947) The Principle of Quantum Mechanics (second edition) Clarendon Press, pg. 12

According to Paul Dirac, the general principle of superposition of quantum mechanics applies to the states [that are theoretically possible without mutual interference or contradiction] ... of any one dynamical system.

We shall proceed by using this definition to properly explain the fundamental different between the spin angular momentum and spin orientation which makes up the spin state of a quantum object.

What measurement means in quantum mechanics

When these words *observation* and *measurement* are used in quantum mechanics, they mean any physical influence that can cause a collapse of the wave function into a definite state. A magnetic detector is often used for measurements because most quantum objects have a magnetic moment. During a measurement, the probabilistic wave function, such as the spin orientation or the position of a quantum object, collapses into a known state. The collapse of the angular momentum during measurement is caused by a well-known phenomenon in the uncertainty principle. Due to the influence from the object used for measuring, the position or spin orientation of a quantum system cannot be found at its current point at the time of measurement.

Pauli Exclusion Principle

By Austrian physicist Wolfgang Pauli (1900-1958). Published in 1925.

The Pauli exclusion principle states that two or more identical particles with half-integer spins cannot occupy the same quantum state within a quantum system simultaneously. The simple form of this principle as it relates to fermions means that no two electrons in close proximity can have the same set of quantum numbers. What this means for the electron is that the spin angular momentum of two electrons emerging from a single source, or two electrons in close contiguity, will be opposite at time $t = 0$.

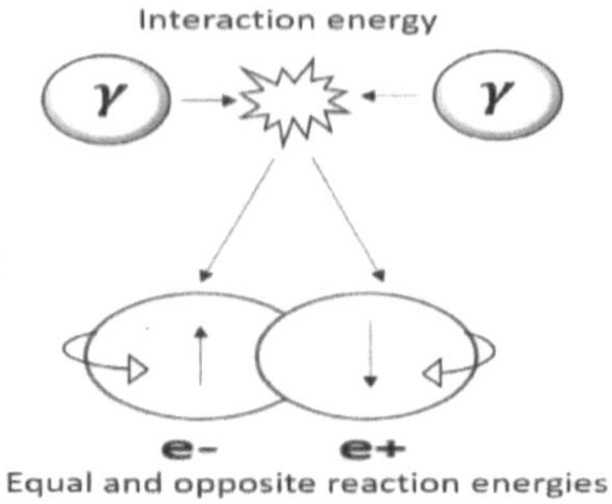

Fig. 301

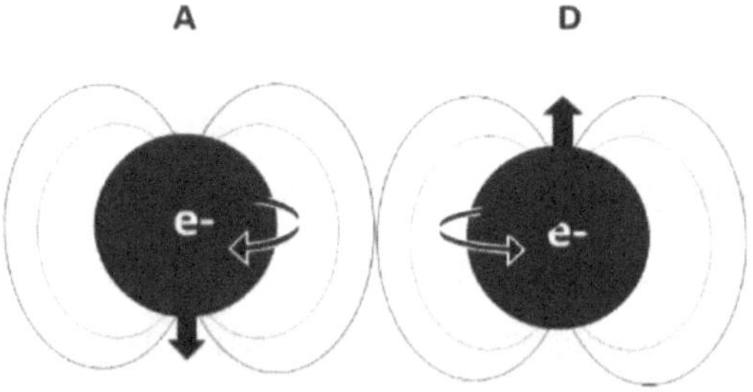

Fig. 302

Fig. 301 shows the moment the collision of two photons becomes a pair of electron and positron. The two arrows represent the opposite spin states of the two electrons at time $t = 0$, as explained in the Pauli Exclusion Principle. Fig. 302 shows another two electrons, A and D, with opposite spin states after interacting via their electromagnetic force at time $t = 0$.

Formulation 1

The wave function (ψ)

The wave function was coined by Edwin Schrödinger and used in his equation to describe the behavior of quantum objects, such as the behavior of an electron in a single hydrogen atom. In quantum mechanics, where the momentum and position of a wave-particle is unpredictable until a measurement is made, the wave function describes the probability of finding a particle in a particular location at a particular time. Also included in the wave function are all the measurable information about a particle.

Quantum mechanics describes the fundamental cause between two entangled particles as sharing *a single wave function*. However, the description of the wave function doesn't elaborate deeper into the reality of the so-called "single wave function." In this formulation, we shall explain and identify a *fundamental variable* in the wave function that allows a correlation between entangled particles.

The spin state of a quantum object, as we shall

explain, is composed of two realities that only accurately commute at the time of measurement. Contrary to a thought experiment by Erwin Schrodinger popularly known as Schrodinger's cat experiment, consider a chameleon laying undisturbed in a dark room, for example. One state could be when the chameleon is green, and another state could be when the chameleon is brown [00.1]. Due to a DNA defect, the chameleon's color is conserved; that is, the chameleon can only exhibit the two colors mentioned here [00.2]. The brain of the chameleon only retains the skin color when it last sensed light [003]. The skin of the chameleon can exhibit any of the two colors or no color when it is not sensing light [00.4]. The chameleon, however, will instantly change its color to the opposite of its last retained color when it senses light [00.5].

Let's assumed that these above details about the chameleon are not known to anyone including quantum mechanics experts and Albert Einstein. Imagine if the only way to know the state of the chameleon was by looking at the skin after we shine a light on it. But before we turn on our flashlight, we ask ourselves what state the chameleon is in under darkness? According to the current main stream argument which started precisely in the year 1927 when Einstein and Bohr began their argument on quantum theory, the mathematics of quantum mechanics tells us to assume that the chameleon is green and brown simultaneously. Albert Einstein, on the other hand, argues that if the chameleon is green when we shine a light on it, it means the color of the chameleon is real, meaning the chameleon was green the whole time while in the dark, and vice versa. As both arguments seem logical if one who doesn't know the above details attempts to predict the state of the chameleon before introducing light into the room, the misconception in both quantum mechanics and the EPR paradox is how they view the *state* of the chameleon.

The two attributes that make up the state of this chameleon is the brain and the skin. In fact, the brain of the chameleon can only sense light or does not sense light one at a time, and the brain always retains the memory of the last single color of the chameleon. Let's assume we finally turn on the light. Using the above details, if the chameleon is green when we look at the skin, even though the chameleon could have been in a multicolor state while lying in the dark, the color green that we observe tells us that the color of the chameleon when it last sensed light was brown. Same thing if the color of the chameleon is brown when we look at the skin.

Take note of four things in this example: (1) when we can't see the chameleon, that is, when the light is not on, the color of the chameleon is not definite; (2) at any given time, the brain of the chameleon always retains the memory of its last single color when it last sensed light; (3) we can never see the current or last single color of the chameleon at the moment we shine a light on it; and (4) the single color we see when we shine a light on the chameleon only tells us the last single color of the chameleon.

Let's bring this example into quantum mechanics. The two attributes that make up the spin state of a quantum object are: (1) spin angular momentum, and (2) spin orientation. Now, compare the chameleon's color to spin orientation, compare the chameleon's memory (brain) to spin angular momentum, and compare the skin of the chameleon under darkness to the superposition of states. The flash light represents a measuring device, while a chameleon in the dark represents an isolated wave-particle, such as the electron.

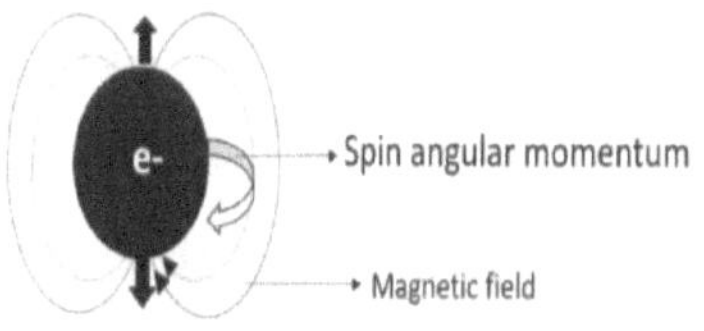

Fig. 742

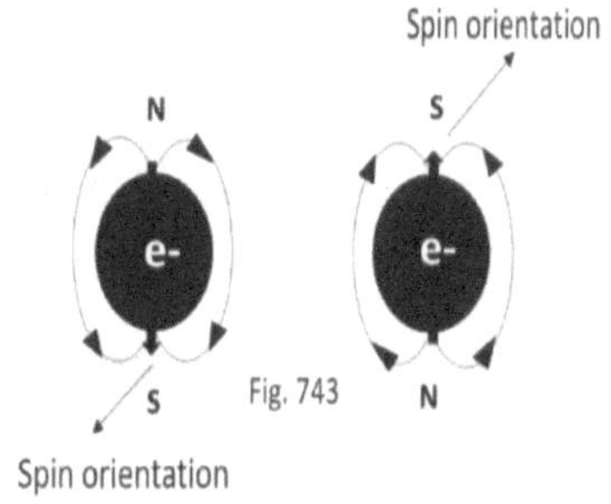

Fig. 743

The spin orientation of a quantum object is a momentary effect of the current spin angular momentum's vector at the time of measurement. *In all the diagrams presented herein, the quantum objects are in a waveform if it has more than one spin orientation. Since we cannot reasonably illustrate the spin orientation of a wave, we illustrate the superposition of spin orientation with the particle form of a quantum object* [23.3].

In Fig 742, the current vector of the spin angular momentum indicates where the spin orientation of the isolated electron's last collapse, and not necessarily what will later be the definitive spin orientation of the electron at the time of measurement [19].

For instance, the spin orientation of the electron in Fig 742 was last detected in the spin down state. On the other hand, until a *measurement* is made, we cannot say with 100% certainty where the magnetic field dipole of the electron is located after measurement. Here, we have made a distinction between the vector of the spin angular momentum and the spin orientation. Take note of the emphasis placed on "measurement," which will be addressed in the next topic.

Like magnets, the electron, like many quantum particles, have a north and south dipole that indicates their orientation in space. However, unlike Fig 743, which carries a North and South Pole symbol, an isolated quantum wave does not carry orientation identifier in reality. During measurement, the orientation of the magnetic field dipole is how we identify the orientation of the electron. For this to happen, the quantum wave has to collapse into its particle form.

Contrary to popular belief, during and after measurement, the system that is definitely in one state is the vector of the spin angular momentum, not the spin orientation, see [19]. The spin orientation of the electron cannot be assigned a definite vector until another measurement is made. Even when a measurement is made, and the spin orientation of the electron is pin-pointed in a particular direction, we must consider the spin orientation to be in that direction *only* at the time of measurement. In this sense, *only* the spin angular momentum of the electron is what collapses (change direction at $t = 0$) into a definite direction during measurement. The spin orientation only collapses momentarily during measurement.

This concept can be understood from a classical perspective. Consider the spin angular momentum and spin orientation of the earth, for example. The clockwise spin angular momentum of the earth is conserved, meaning that it neither spins faster nor slower. However, the magnetic field of the earth periodically changes its dipole. In this example, you can see how the vector of the spin angular momentum is definite in one state, while the magnetic field dipole changes direction. Due to the earth being solid, the spin orientation of the earth doesn't change when the magnetic dipole switches. The angular momentum of the electron behave quite similarly to that of the earth, but different when it comes to spin orientation. The difference lies in the fact that the electron is a pure fuzzy energy or wave when it is not disturbed. Thus, when the electron's spin angular momentum is definite in one direction until it experience a force, the spin orientation of an isolated electron wave evolves with time when it is not experiencing a force.

In summary, only a *measurement* (external force) can collapse (change); in this case, the vector of the spin angular momentum. It is only at the moment when the spin angular momentum changes vector that we know the spin orientation is definite in one place. Again, bear in mind that we placed an emphasis on "measurement" in this paragraph.

Before Pauli exclusion principle at $t=1$

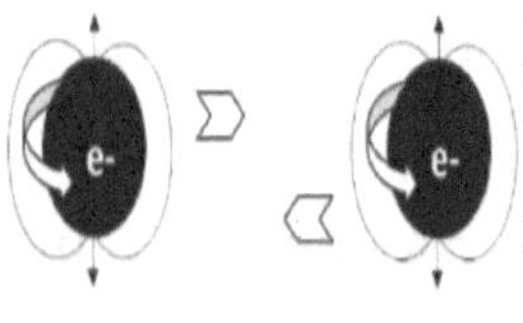

Fig. 091

Pauli exclusion principle at $t=0$

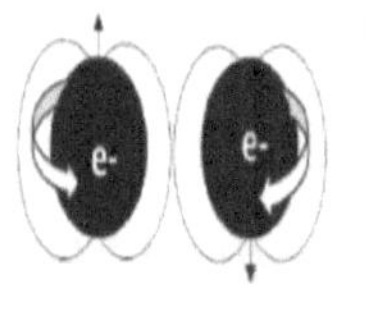

Fig. 092

After Pauli exclusion principle $t+1$

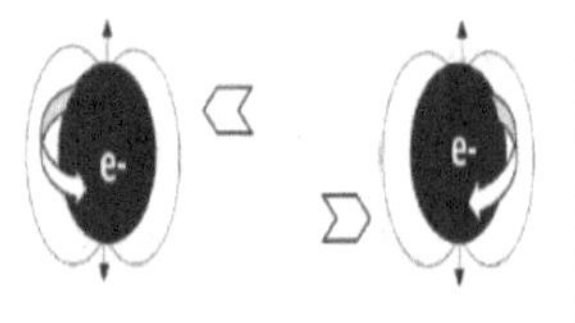

Fig. 093

In Fig. 091, despite the superposition of the spin orientation, the vectors of the angular momentum indicate that the last collapse experienced by both isolated electron waves [23.3] was in the spin up states. Fig. 092 shows the moment both electron waves in Fig. 091 interact, with their angular momentum collapsing into opposite vectors, thus causing a magnetic moment in the spin orientation of the electrons in space. As a result, for the electron on the left, at $t = 0$, we know the orientation of the electron in space was at the spin up state, and the electron on the right was at the spin down state. At this time $t = 0$, we assert that no superposition of states occurs within the electron. However, at time t + 1 (Fig. 093), after both electrons have separated, even though the vectors of their angular momentum still indicate specific states for both particles, for an isolated electron wave, meaning an electron not under the influence of any force, the spin orientation is in a superposition of state. In other words, there is no external magnetic force to give the electron a magnetic moment. In fact, in the absence of external force, the electron can only be described as a wave.

We learned in classical physics that matter primarily exists in liquid, solid, gaseous, and most recently, plasma states. However, given the various observations noted during quantum experiments, like the famous double slit experiment by Thomas Young, we know that matter can also exist as a wave. An electron, for instance, does not lose its mass when it is in form of a wave. Thus, based on the definition of matter, a quantum wave must be considered not only as a type of motion but also as a quantum state of matter. Obviously, a wave has no definite orientation or position in space. As a result, because of the dimensional freedom associated with waves, we must also concede that the reality that allows the mathematical description of the superposition of a spin state is the wave form of a quantum object.

A measurement is known to collapse a quantum object from a wave-particle duality into a particle state. Since a quantum system can only yield accurate information when it is in a particle state, we know the latter assertion is true because when a wave function collapses, it yields definitive information about its position or spin orientation. Moreover, quantum mechanics only refers to the collapse of a wave, not a particle. As a result, you can see why a quantum wave is said to have no definite spin orientation until it is subjected to an external force.

During the Pauli exclusion principle, as seen in Fig. 092, because the isolated electrons choose the spin up and spin down states freely, they are therefore considered a natural reset orientation of the electron's spin. In quantum mechanics, these spin states are said to be quantized. But when a non-quantum measuring device is imposed on the electron, the angle where the spin of the electron will be oriented in space depends on the axis chosen by the measuring device.

Fig. 317

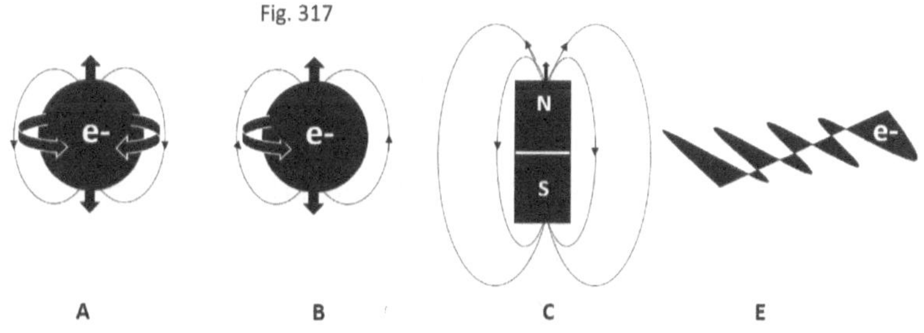

The purpose of Fig. 317 is to show the differences that suggest the superposition of state for the electron. In addition, it shows the physical reality of an isolated electron wave. Diagram C shows a solid isolated magnet with one orientation. The magnet is made up of many charged electrons with neighboring hadrons. At any given time, the magnet can only assume one spin orientation. On the other hand, the electron, despite having mass, is simply a purely indivisible charged wave. When the electron is not disturbed, it evolves into a wave, as seen in diagram E. In the wave state, a quantum particle has no reasonable spin orientation. The intrinsic spin angular momentum of diagram E, however, remains unchanged until the wave is disturbed by external force.

The spin angular momentum vectors in diagram A is not theoretically possible, and the spin orientations are a theoretical illustration of uncertainty in orientation when the particle is in its waveform. This uncertainty could be described as all possible orientations or none. Diagram B is an acceptable illustration of an undisturbed (isolated) electron [23.3]. The clockwise spin angular momentum of B is an indication that the electron's last collapse will be in a spin up state. However, after measurement, the spin orientation is not definitive in any axis, as seen in diagram E, until it is subject to another measurement.

Conclusively, because the spin angular momentum is an important cause behind magnetic field, we cannot always define magnetic moment as an intrinsic attribute, *except* at the instant when the quantum wave is subjected to an external force such as measurement. This is especially true when studying the spin angular momentum of neutrinos, which have a spin angular momentum but no electric charge. Consequently, the spin angular momentum is an inherent property which, like charge or mass, is real at all times. However, the spin orientation of a wave-particle, which is mostly described by the magnetic field dipole, is only real at the time of measurement (magnetic moment).

Single wave function $|\psi|\ \alpha\ |\psi|$

Consider a common assertion by most physicists, which often goes something like: “entangled particles stayed connected even if they are separated by light years” [66.6]. To begin with, we can all agree that no one has ever measured entangled particles that are located light years apart. Thus, from our knowledge of how theories are written in physics, quantum mechanics is indirectly admitting the existence of a fundamental variable within the wave function, which must remain undisturbed for entanglement to be proven over a long distance. We know this is true because it is also the assertion of quantum mechanics that entanglement is destroyed once the wave function collapses or is disturbed by an external force. In theory, what this assertion is saying is that after two particles interact or separate from a single source, provided a certain variable in the wave function is not disturbed, a strong correlation will exist between the two particles, regardless of the distance between them. Based on all the explanations we've presented thus far, it is our assertion that this fundamental variable is the spin angular momentum.

When two physical objects are not connected by matter or energy, they are separated by distance in space. Otherwise, contrary to popular view, the "single wave function" shared by entangled particles are not in any way connected by a quantum field or any physical means. Rather, the reality allowing the correlation of spin states between two entangled particles are two identical yet independent wave functions with opposite vectors that describe the current state of each spin angular momentum at any given time when the system is undisturbed, but are subjected to instantaneous changes and a definite but temporary spin orientation when influenced by an external force.

Measuring the wave function ψ

To obtain information about a quantum system, the quantum object must be measured. It is a well-known fact in quantum mechanics that any measurement, whether intentional or unintended, made on a quantum object, will immediately collapse the system into either a specific position or a specific spin state. The latter depends on the value that acts upon ψ.

That being said, this is how quantum entanglement between electrons works in theory. We've assumed that the measurement imposed on two independent electron waves $|\psi|$, $|\psi|$ was designed to yield a value corresponding to the spin orientation of the electrons. We further assumed that the two independent ψ, which are our two quantum systems, had previously interacted. Prior to measurement, we are uncertain of which direction the angular momentum of both electron waves are pointing. However, due to the Pauli exclusion principle, provided the waves hasn't collapse at a time unknown, we are certain that the angular momentum of both electrons has an opposite vector at $t = 0$ of interaction.

The only way to know for certain which electron has the angular momentum pointing left and vice versa is to subject one or both electron waves to measurement. However, the moment both waves are measured, the vectors of each angular momentum will experience a force that will immediately collapse them into opposite directions. By observing what is revealed by the spin orientation during measurement, we then know which electron has the vector pointing left or right *prior* to measurement. Despite that the new vector of the angular momentum stays the same until another force acts on it, the *spin orientation* that we detect at the time of measurement $t = 0$ cannot be considered the actual spin orientation of the electron after measurement. In this case, the spin orientation is only determined at the time of measurement and in the angle (axis) imposed on the electron by the measuring device.

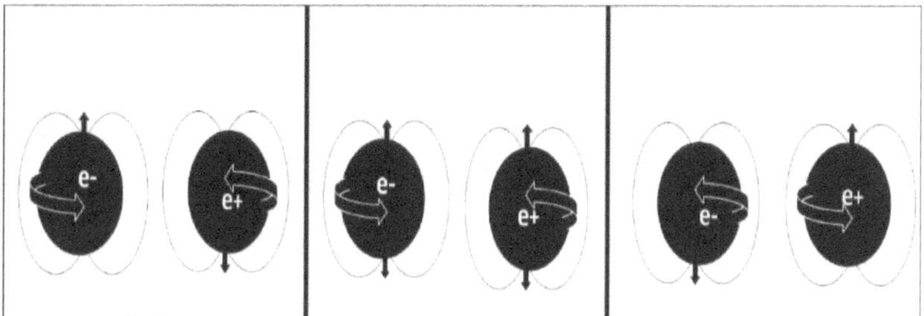

Fig. 050 Fig. 051 Fig. 052

Fig 050 is what occurs at the *instant* when a spin zero particle becomes two half spin particles. Unfortunately, contrary to what EPR suggests, Fig. 051 [23.3] shows what happens after the entangled particles separate. Fig. 052 shows what happens to the angular momentums in Figure 050 when they are later measured at $t = 0$.

In conclusion, due to the uncertainty principle, the angular momentum vectors in Fig.050 will not be found in the same exact state when they are measured later. In addition, during measurement, the angle of spin orientations depend on the axis chosen by the measuring device.

If any of the electrons in Fig.050 are measured later and the angular momentum vector is found in the same state, it means, at a time unknown, the electron has been influenced by external force prior to measurement. Any argument against this can easily be defeated with the wave function collapse problem, which we shall discuss later. If not, there is no logical reason why the spin angular momentum vectors in Fig. 050 should remain the same during another measurement. The probabilistic results, which we often claim to be the nature of quantum mechanics, as it relates to spin orientation, is due to the possibility of the wave function to collapse at a time unknown. Thus, one cannot predict with certainty what the result of the measurement will produce. Otherwise, if the wave function remains undisturbed after its current state is known, we can predict with a high degree of certainty what the spin orientation of a particle will yield in the future when it is measured. This high degree of certainty is often seen when measuring two entangled particles immediately after they separate or in a controlled environment. Usually, most entangled particles are measured immediately or at a short distance after they separate from a single source. The probability that a quantum system will come into contact with an external force is lower if it is measured immediately or at a short distance after emerging from a source.

Equal time, and time delay in entanglement (11.1)

Seconds are often used to measure changes in classical mechanics. However, within a second, there are other smaller units of time, such as microseconds, nanoseconds, picoseconds, femtoseconds, and Planck time. Because of the smaller units of time, when using seconds to record time in quantum experiments, the equal time may vary.

Out of curiosity, you may have considered whether the unit of time we use during a quantum experiment matters. Using the speed of light as an example, you will understand how equal time could vary in other smaller units of time, and yet still be equal to a second. On July 4th, 1977, at exactly 3:16:00 AM, a physicist at point *A* sent out two photons *D* and *P* simultaneously to two detectors. Detector 1, where Photon *D* is heading, is located 46,500 miles away, and detector 2, where Photon *P* is heading, is located 93,000 miles away. On the same day, July 4th 1977, when the results from both detectors were revealed, the arrival time for photon *D* as recorded by detector 1 in nanoseconds was 3:16:0.000000025 AM, and the arrival time for photon *P* as recorded by detector 2 in nanoseconds was 3:16:0.000000051 AM. However, when the results recorded by the same detectors on the same day in seconds were revealed, the arrival time for photon *D* as recorded by detector 1 was 3:16:01 AM, and the arrival time for photon *P* as recorded by detector 2 was 3:16:01 AM.

It is apparent to anyone that there is a 46,500-mile gap between 46,500 and 93,000 miles. However, because a photon can cover a distance of more than 139,500 miles in a second, the difference in distance does not count when measuring in seconds. In fact, the photon can cover a distance of 186,000 miles per second. Similarly, during an experiment involving quantum particles, any smaller unit of time difference between the properties of two identical particles will correspond to an equal time when measured with seconds.

In conclusion, as explained by photon *D* and *P*, one who is using the distance between detectors to create a time delay should take into consideration the speed of light. For example, to delay the a photon coming out of point *A* for 1 second, detector *B* has to be located more than 187,000 miles away, not within a physics laboratory.

Using time to disprove information transfer (11.2)

Like [66.6], there are many claims in quantum mechanics which many people believe is true but in reality they have no testable prove. In some cases, no experiment can be conducted to test most quantum phenomenon. We simply need to depend on reasoning to draw our conclusions. For this reason, our approach to understanding how quantum entanglement works must include a rigorous scrutiny of every claim associated with entanglement.

One of the reasons why we shall further elaborate on the quantum entanglement theory in this chapter is because of a popular view in quantum theory that an undisturbed particle collapses at the *instant* when the other entangled pair is measured [65.5]. This assertion only opens the possibility that either quantum entanglement happens because of simultaneous measurement or via information transfer.

The legitimacy of this view is dependent upon knowledge, which, in quantum mechanics, is subject to measurement or observation. Therefore, anyone who agrees with this view should consider this important question; what test can we perform to show that an isolated entangled particle collapse at the *same instant* that we measured its entangled pair? This is equivalent to claiming a chameleon *M* inside a dark room changes color at the *same instant* when someone shines light on his sibling *N* at another location. Yet, we all agree that 1, the act of shining light on a chameleon changes the chameleon's color instantly, and 2, the only way to know if a chameleon changes color is by shining light on the chameleon. If one claims that the act of shining light on chameleon *N* is what changes the color of chameleon *M*, yet, the only way to prove that the color of *M* changes is by shining light on it, the argument one must first overcome here, before a reasonable claim can be established, is to disprove that the act of shining light on chameleon *M* is not what changes its color.

The bedrock of the uncertainty principle is rooted in the effect of measurement on a quantum object.

For instance, the five senses of a human cannot observe, detect, or interact with anything in space without involving the exchange of energy or matter. In this instance, quantum objects can be compared to human senses. A quantum object needs itself or the exchange of another quantum object to interact with the world around it.

The problem we will encounter if we agree with the popular view of quantum mechanics [65.5] is often referred to as the wave function collapse or measurement problem. How do we know if a quantum wave has or has not collapse after the last known collapse occurred? In other words, how do we obtain or retain the information about a quantum system without collapsing its wave function? This unfortunate act of acquiring knowledge is analogous to a person who only speaks English using the English language to ask an audience in a room if there's anyone here who doesn't understand English, and he or she expect those who doesn't understand English to raise up their hands. This person will never know those in the audience who does and doesn't understand English, because, those who understand English will not raise up their hands, and those who doesn't understand English will not raise up their hands.

Measurement is the only language we know that can give us information about the state or position of a quantum object. Therefore, to test for chameleon *M*, we need a detector to measure the exact time that an entangled object receives the faster than light information. To do so, we need the detector to record particle B at the same instant when we measure particle A without disturbing particle B. But the detector will eventually collapse particle B at the same time when we measure particle A. Thus, we cannot do a test to prove information transfer, or the instantaneous collapse without validating the concept of equal time measurement. On the other hand, unless we can prove the existence of the faster than light mystery particle delivering information to particle B, we can assert the quantum entanglement theory proposed herein, and [65.5] would still remain an unproven hypothesis.

So, here is the underlying logic. Information transfer and [65.5] are mere suggestions, not the prediction of any solid theory pending validation. Nonlocalism in quantum mechanics only suggests that entangled particles have a stronger correlation and do not necessarily send or receive information. In other words, information transfer or the instantaneous collapse of an undisturbed particle is an unproven view of quantum mechanics.

However, we shall now examine the possibility of information transfer. We shall see if, in theory, a mystery particle could be responsible for conveying information between entangled particles.

Let our mystery particle be real; that is, the nature of its existence does not theoretically violate the laws of physics, or present a logic that leaves a reasonable doubt about its existence.

Taking the electron as an example, in quantum physics, the information transfer between two fermions is accomplished by a gauge boson. Thus, we can easily conclude that our mystery particle must be a boson.

In the quantum entanglement hypothesis, the mystery boson is assumed to move faster than the speed of light. Hence, we can also conclude that for our mystery boson to be able to move at or faster than the speed of light, it must be massless.

Since our mystery boson is massless, it cannot belong in the weak force category. Additionally, since strong forces are only effective within a short range, we can conclude that our mystery boson is not in the strong force category either. Massless bosons do not have an electric charge. For as far as we know, a body can have mass and lack an electric charge, but anything that has an electric charge must have mass. Thus, we can further conclude that our mystery boson has no electric charge.

As we've agreed in previous claims, we can agree that our mystery boson must be stable and must have at least one intrinsic property such as spin in other to be able to hold information during the course of travel. Here is where the existence of this mystery boson becomes impossible. It is the hypothetical claim of quantum mechanics that entangled particles are affected instantaneously regardless of the distance between them. This in theory means, during measurement, if we have two entangled particles located 14 billion years apart (corresponding to the postulated age of the universe) and they are still moving apart with expanding space, our mystery boson can reach them at infinity instantaneously without experiencing time. This also means that the mystery boson needs infinite energy to accomplish this. Lastly, since entanglement is always guaranteed after two particles interacted, it means the wave function of the mystery boson cannot collapse under any circumstances. If any physicist needs additional prove to disprove the existence of this mystery boson, such person will have to invent a new definition of prove that he or she requires. As far as our understanding of the universe is concerned, in theory, this mystery boson cannot possibly exist. As a result, information transfer is not a cause behind instantaneous or delayed correlation of entangled objects.

With this conclusion, we are left with two possible cause behind the correlation of entangled particles: (1) that the vector of angular momentum, or in the case of the photon, the polarization vector, collapses during measurement and remains unchanged after measurement until another force acts on it. Or (2), simultaneous change and simultaneous time is responsible for quantum entanglement.

Predictions for non-localism

As shown in (1.22), we can easily prove entanglement in classical mechanics and there we cannot but agree with locality. This test has been done repeatedly, and it works as the theory predicted. Anyone who doubts it is free to put it to test.

Frankly speaking, assumptions, beliefs, or one's opinion are irrelevant when prove has been presented. But without a solid logical prove of non-localism, these are two immediate problems that quantum mechanics should expect if it accepts non-locality.

1. The first problem has to do with philosophy. We can imagine a day that a prosecutor will claim that a punch thrown by someone located 3,000 miles away from point B led to the injury of someone at point B. According to quantum mechanics, an expert witness in this area of study must affirm that a similar action to this statement is possible in physics. However, one thing is certain, locality will win this case in a

court of law, especially when quantum mechanics explain the physics behind non-locality.

2. The second problem has to do with reason or intent. This problem can be summarized with a simple question: why is it important for an inanimate object to inform another inanimate object about its current state during measurement? For particles in close proximity, the reason for the Pauli exclusion principle is reasonable and obvious. But when two particles are separated by several distance, the Pauli exclusion principle no longer applies. Thus, any intentional action taken by an inanimate object without a logical explanation is unscientific. Except in religious beliefs, any tangible cause and effect that cannot be explained within the concepts of space, energy, matter, and time is a delusion.

If quantum mechanics, however, decides to stick to its assertion of non-localism, it means we have only presented a classical entanglement theory. But, in a world where science is established on the principle of empirical, testable, and demonstrable protocol, the very bases of reasoning, if science believes two inanimate quantum particles that are separated by vast distance can have instantaneous correlation with intent, we can thus conclude that science has found proof of the supernatural.

Bell's theory and tests

J. S. Bell. [ON THE EINSTEIN PODOLSKY ROSEN PARADOX]
Physics publication Co.
(1964) Vol. 1 No. 3 Pg. 195-290

The argument of both EPR and quantum mechanics was based on spin orientation, in which if one looks at it based on the explanations in this chapter, EPR is wrong and quantum mechanics is right on predetermined state. On the contrary, based on the description we presented on the spin angular momentum, EPR is right and quantum mechanics is wrong on localism. The Bell's inequality approached the argument with a logic that is based on spin orientation.

When inequalities is demonstrated with mathematics or a classical method, the method involve uses the manipulations classical and tangible things. This perfectly commutes with the description (not definite until measurement is done) of the spin orientation. Thus, if the Bell's inequality is used in physical experiment to test the EPR's view of predetermined state, which in theory is said to be real regardless of any manipulations, the EPR's view will be found to be wrong.

In fact, Albert Einstein did not use the words "hidden variables" in the EPR paper, nor did the publication provide a specific definition that described the supposed additional variables. If additional variables are intended to mean "a predetermined agreement made between two entangled particles which they must always adhere to regardless of the axis of any measuring device or time of measurement," then additional variables do not exist in this sense. However, based on the intent of the EPR argument, in which the main point was to restore localism in quantum mechanics, I would argue that the term hidden variables is misleading when used in the absence of the full knowledge involved in the workings of quantum entanglement. The fundamental variable which we have used to replace this word means "any mechanics or variable within two entangled particles which allows the result of their spin states to correlate when measured."

As it seems, contrary to the EPR suggestion that measurement reveals predetermined states, in fact, the spin orientation of an entangled particle is only known at the time of measurement, and the axis in which the spin state will be oriented can be affected by the measuring device.

The conclusion in Bell's formulation (p. 199), showed that Bell's theorem intend to resolve the issue of additional variables by manipulating the settings of the measuring device used to measure entangled particles. This only serves to disprove the notion that spin orientation of entangled particles are predetermined, meaning they will be found in their current state and specific axis at the time of measurement.

High probability of quantum states

Since we are limited to one outcome out of two spin states when measuring a quantum object along a specific axis, there is a 50% probability of getting entanglement results when measuring two non-entangled particles. Take two electrons, *x* and *y*, for example. At different laboratories, two physicists measured the spin orientations of two non-entangled electrons that were sent to them anonymously. Both physicists did not know that the electrons were not entangled. Electron *x* was measured and the spin orientation was spin up, and at the same time in another location, electron *y* was measured and found to have a spin down orientation. When the two physicists compared their results, they concluded the electrons were entangled. We know these two electrons were not entangled. The two physicists only got lucky. One way to avoid this kind of error in judgment is by repeating experiments multiple times.

Prediction

Fig. 316

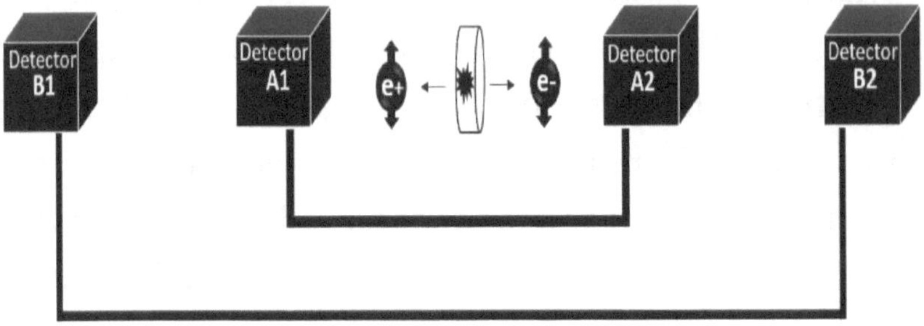

If the mechanics of entanglement we put forward for the electron is true, such that the spin angular momentum unlike the spin orientation remains unchanged until during measurement, after two entangled electrons are measured, provided the spin angular momentum remains in its collapse vector, the entanglement between them will continue.

Thus, consider two entangled electrons that are measured at detectors A1 and A2, whose angular momentums collapse in opposite directions. According to quantum mechanics, entanglement is destroyed after the measurement is taken. However, according to the quantum entanglement theory described herein, since the mechanics in the Pauli exclusion principle that led to entanglement is the collapse of two spin angular momentum in opposite directions, because the spin angular momentum of the electrons in Fig. 316 collapses in opposite directions at detectors A1 and A2, both electrons will be entangled when measured again at detectors B1 and detector B2.

For entangled photons, since the polarization of both photons collapses at detector A1 and A2, both photons will be entangled when measured again at detector B1 and B2.

Chapter 9

The Universe as God: The Third Law of Existence

I must confess, the more I study the universe, the more it seems that it is an intricate equation and I'm just here to solve it. I hope I haven't failed to include something crucial in this first publication of the multiplication theory, as it is still in its infancy and I'm still trying to understand the language of mathematics. But what good will it do me if I keep the little I know to myself simply because I can't speak an abstract language like a mathematician? One could keep a secret forever, but wisdom is like a guilty conscience: it won't let you rest until you share it.

The quest that led to the ideas you've read in this book arise from a question I've been asking since I was a teenager: What is God and everything? The way one approaches this question matters. If you believe the answer to this question is complicated, you will never recognize the answer even if you see it. On the other hand, if you approach it with a simple and open mind, the meaning of God and everything will become clear to you.

This doesn't mean that when you discover the true meaning of God, you will know and understand every phenomenon in existence. Perhaps this next wisdom of mine would help you understand what I meant when I said, "Everything will become clear to you: God has a secret hidden in plain sight. The problem is that the more you learn about the secret, the more complicated it becomes."

I have had dreams that I can't explain. I have seen the might of the ocean from up above. I have stared outside the earth and beheld the glory of planets and stars. Yet nothing I have seen or heard comes close to the existence of one, an entity so enormous it makes the speed of light slower than the crawl of a snail. Its height, length, and width are immeasurable. It cannot be touched or smelled, but despite this, nothing can hide from its presence. We can't see its physical appearance, yet it is more real than the things we can see.

None of the words invented by man are enough to encapsulate its majesty. Nothing within it is big enough to describe its glory. The might of its wisdom shines like daylight in the clusters of galaxies, just as it can be seen clearly in the atoms that make up the physical world. It is the author of life and death, the architect of stars and planets, and the breath of life. And whether we like it or not, we can't escape its presence, not even in death.

If this entity sounds familiar, before modern civilization, we called it God and worshiped it. In this age and time, the majority of people prefer to call it the universe and study its vastness.

When someone asks me if I believe in God, I usually ask them to define the god in question. The definition most people present often suggests they are referring to the image of a supernatural deity.

As I have come to realize, there are two Gods: one can be *proven* to exist, and the other is imaginary. The God I believe in is the God whose state of existence is an everlasting occupation that has no beginning or end. Out of his invisible nature, all visible matter was made.

The characteristic of this everlasting occupation is three, which I have clearly seen in all visible matter. Thus, to me, belief and faith are not blindly accepting a state of existence without evidence. Instead, they are accepting a state of existence I already know contains more reality than any of my five senses can interpret.

The other god I don't believe in is the Homo-sapiens-looking infinite god who created an infinite universe. This god only exists in the minds of humans who have failed to apply wisdom to their religious studies.

I must confess, the Holy Bible is the greatest historical text I've ever come across. This holy book, though written before modern civilization, contains numerous codes about most discoveries we are uncovering today with modern science.

Spirituality, intelligence, and religion

There's a difference between being spiritual and being religious. To me, being spiritual means one acknowledges that there's more wisdom, knowledge, and understanding to the existence of the universe than the five senses of life-forms could explain. This humble acknowledgment by many people has led to unproven beliefs, such as religion.

I believe dream is the fifth dimension, and I believe we are not alone in the universe. If you think these beliefs make me absurd, then I can equally say you are a fool in your noble, sophisticated, and moral beliefs.

Imagine having a dream where you saw your dog involved in a car accident. In the dream, you got pulled over by the police on your way to the veterinary hospital, and in the hospital parking lot, you had a flat tire.

A week after you had this dream, your dog actually was involved in a car accident, and on your way to the hospital, you were pulled over by a police officer, and then you had a flat tire in the hospital parking lot. As you wonder about this correlation, human arrogance might convince you to see this as just mere coincidence, or you can humbly agree with your innermost conscience telling you there's more to reality than your five senses can help you understand.

One doesn't have to belong to a religious group to validate one's morality. What is not factual is when a human denies the universe of its accomplishment. To deny spirituality is to deny that consciousness is something that emerges from the universe. Despite all the methods humankind has deployed to discredit an intelligent universe, at the end, life-forms are simply conscious quarks and thinking electrons. Can anyone deny this?

Accordingly, to deny an intelligent universe is to deny oneself. Because unless elementary particles, as we refer to them, have intelligent ability, in a trillion years, they would never evolve into life-forms. Perhaps the question now should be: What are quarks, bosons, and leptons before we assign those scientific names to them?

What is this?

In the introduction, I asked a question twice: "What is this?" I knew you wouldn't be able to answer the question if you couldn't identify those symbols. I used this wisdom to remind you of our humble beginning during the prehistoric age, when all we could do was stare at everything and wonder what it was. So before you accuse me of using scientific names to describe God and everything in existence, I hope this will remind you that the universe and everything in it are not made by science. The words *energy*, *matter*, and *space* are just logical names we invented to describe the characteristic nature of something that exists.

I am what I say I am

I stated in the introduction that not everyone will agree with me. I understand if you happen to be one of them. Occasionally, when I consider the rate at which I think, I wonder if I'm so foolish that I'm beginning to think I'm wise, *or* I'm so wise that I'm beginning to think I'm foolish. Am I wrong, or is the universe just obsessed with three things? Did the earth favor life because it's the third planet from the sun?

Paradoxically speaking, perhaps you are the smart or intelligent one, and I am the wise monkey. But with all your prestigious education and higher cognition, you can't deny my existence as part of the universe. In fact, I am an intelligent representation of the universe—and before you give all kinds of scientific and religious names to identify me, I am just what I am.

Since I was a kid, I've searched for evidence of God's existence. All the while, I didn't know I was looking for myself. Now that I've gotten rid of all the worldly nomenclature, I've realized *I am that I am*. I am the alpha, and I am the omega. I am the beginning, and I am the end. I can change from one form to another, yet what I'm made of can neither be created nor destroyed. My wisdom is not infinite; neither am I omnipresent. But make no mistake. When you see me, you see God. Likewise, when you see God, you see me. And the same is true for you reading this book, for it is written, "I have said, Ye are gods; and all of you are children of the most high" (Psalm 82:6).

Purpose of life

With everything that humans have discovered and accomplished, mankind doesn't need the discovery of a new reality to understand the things they don't understand. The sooner we begin to think critically about what we have learned and seen in reality, the sooner we will be relieved from our mysteries.

A sophisticated problem requires a sophisticated solution. Sophisticated solutions, however, are not built directly with what one has learned, but by how well one can manipulate that which they have learned.

Life-forms are not here to sleep, eat, and play alone. These things are just necessary for survival. It is beyond coincidence that every tangible form of existence that came out of the intangible fabric of the universe has the ability to change from one form to another, either by way of decay, reproduction, or multiplication. With this, it could be said that therefore, one of the fundamental principles of the universe is to ensure that existence continues, *and* continues to improve.

Take a rich person for example. Almost every rich person want his or her wealth to continue or increase, even after they are dead. That's why they invest, and start businesses here and there. The same could be said about the universe. The purpose of creating something with the ability to change, multiply, or reproduce is to ensure that existence continues. For this reason, all created things must have the ability to multiply, reproduce, increase in number, or change from one form to another. This fundamental principle is the purpose of life. Or as it is written in Genesis 1:28, "Be fruitful, and multiply, and replenish the earth, and subdue it."

Think about it: if the purpose of life and freedom is for life-forms to roam the world, this begs the question of why life formed in the first place. If a creature doesn't have its own purpose in existence, you better be sure the creator has one for it. Otherwise, if a creator doesn't have a purpose for its creature, that creator is not intelligent in nature. In this case, a creature is just a mere effect that came into existence as a result of its creator's action.

If I could disprove intelligence in the behavior of the universe, I would argue against a reasonable purpose of life. But I have examined the wisdoms of science that man uses to understand reality, and no matter how much I tried, I can't disprove intelligent universe beyond reasonable doubt. The ability to apply mathematics to solve a problem is an act of intelligence, one that is obvious in all created things.

In the beginning, when the only method the universe used to reproduce was multiplication, it's reasonable to see how particles and bacterium were able to bring forth their own kind by way of multiplying. But considering that the universe prefers to use less energy to do work, it seems as if life-forms will continue to evolve until one that has the ability to create or reproduce its own kind has evolved.

Perhaps the next phase is for life-forms to create another form of intelligent being capable of reproducing or creating another intelligent being. Apparently, it seems as if evolution has stopped at the human being, but still has more work to do to our brain.

Third Law of Existence

If there's a creature, there's a creator—and the purpose of a creature can either be an effect or to serve the creator. Without help or power from the creator, a creature cannot function independently, and it cannot acquire knowledge beyond its ability unless it has been given that knowledge or ability by its creator.

Simplicity

One of the reasons I decided to make this book short and less complex is because I know most people are not interested in what they can't prove or demonstrate. For instance, it would be easy for you to understand that $3 \times 3 = 9$. Most physicists would understand if I claimed that the collision of two protons (that is, two equal but opposite particles containing three fermions each) can bring into existence the most massive fermion known to man. But I'm sure many of you would label me unreasonable if I claimed that by faith, one could use the multiplicative power to multiply five loaves of bread and two fishes until it is enough to feed five thousand people.

As easy as it is to understand the multiplicative power through mathematics, biology, and quantum mechanics, it is almost impossible to explain it with classical mechanics without sounding nonsensical to the casual mind. Thus, I'd say that knowledge is good. Knowledge is power. But not all knowledge can be fed to a hungry mind. Some knowledge is better left in secrecy.

The mind of God

If the universe could speak, the accurate occurrence in space-time, such as force, time, and direction, are how it would talk to us. And if speech does come from the mind, accurate occurrences, such as change and symmetry, are how God would communicate what's on his mind.

Symmetry, as I refer to it, is the language of the anonymous cosmos. This would be our best option to start communicating if we encounter other intelligent beings in the universe. For if an intelligent being sees or observes something happening once, it's an incident. In the same pattern, if the same thing is seen or observed happening again the second time within a short period, it could be considered a random coincidence. But if that same thing happened a third time in the same pattern, in a short period, it's very likely that an intelligent being would *suspect* a reason or intelligent involvement behind it, and thus would investigate the situation.

If you are religious, you can read about a similar situation in the first book of Samuel, chapter three, of the Holy Bible.

In our physical world, we know the past and present, but we are shrouded in the mystery of the future. Imagine if a human knew the speed and direction of every change before it occurred. They would be able to speak confidently of the future—just like God.

Asterism principia

Made in United States
Troutdale, OR
11/03/2026

56844439R00146